Synthesis Lectures on Sustainable Development

This series publishes short books related to sustainable development practices relevant to engineers, technologists, managers, educators, and policy makers. The books are organized around the United Nations Sustainable Development Goals for 2015–2030. Design for sustainability along with the economics of sustainable development will be the common themes for each book. Topics to be covered will span all the major engineering disciplines along with the natural and environmental sciences that contribute to an understanding of sustainable development. The goal of this series is to make theory and research accessible to practitioners working on sustainable development efforts.

Md. Rabiul Islam · Behnaz Behi · Faisel Tubbal ·
Qinghua Guo · Sasha Nikolic · Raad Raad

Universal Access to Sustainable, Affordable, and Reliable Electricity

Critical Insights for the Next Generation of Energy Systems

Springer

Md. Rabiul Islam
Faculty of Engineering and Information Sciences
School of Electrical
Computer and Telecommunications Engineering
University of Wollongong
Wollongong, NSW, Australia

Behnaz Behi
Faculty of Engineering and Information Sciences
School of Electrical
Computer and Telecommunications Engineering
University of Wollongong
Wollongong, NSW, Australia

Faisel Tubbal
Faculty of Engineering and Information Sciences
School of Electrical
Computer and Telecommunications Engineering
University of Wollongong
Wollongong, NSW, Australia

Qinghua Guo
Faculty of Engineering and Information Sciences
School of Electrical
Computer and Telecommunications Engineering
University of Wollongong
Wollongong, NSW, Australia

Sasha Nikolic
Faculty of Engineering and Information Sciences
School of Electrical
Computer and Telecommunications Engineering
University of Wollongong
Wollongong, NSW, Australia

Raad Raad
Faculty of Engineering and Information Sciences
School of Electrical
Computer and Telecommunications Engineering
University of Wollongong
Wollongong, NSW, Australia

ISSN 2637-7675 ISSN 2637-7691 (electronic)
Synthesis Lectures on Sustainable Development
ISBN 978-3-032-19516-6 ISBN 978-3-032-19517-3 (eBook)
https://doi.org/10.1007/978-3-032-19517-3

This Springer imprint is published by the registered company Springer Nature Switzerland AG
The registered company address is: Gewerbestrasse 11, 6330 Cham, Switzerland

Preface

Electricity has become the foundation of modern civilization. It powers homes, industries, transportation systems, education, health services, and communication networks, enabling the growth and prosperity of societies worldwide. Yet, paradoxically, while electricity has never been more essential, access to it remains unequal, and the way it is produced continues to harm the environment. Climate change, fossil fuel dependency, and rising global demand have created an urgent need for rethinking how electricity is generated, delivered, and consumed.

This book arises from that urgency. It presents a framework for what we call "Three-able energy systems", electricity systems that are sustainable, affordable, and reliable. These three qualities are not optional; they are interdependent. Sustainability is vital to reducing emissions and preserving ecosystems. Affordability ensures that access to electricity does not become a burden or a privilege for only a few. Reliability guarantees that electricity is always available at the quality required for daily life, industry, and innovation. Without achieving all three simultaneously, progress toward universal energy access will remain incomplete.

The chapters that follow explore this challenge from multiple perspectives. We begin by examining the global energy situation, emphasizing the pressures of population growth, rapid urbanization, and environmental degradation. The discussion reveals that traditional fossil-based systems, while once sufficient, are now incompatible with long-term sustainability. Renewable technologies—solar, wind, hydro, biomass, geothermal, and even nuclear—offer cleaner pathways, but each comes with technical, social, and economic challenges that require careful consideration.

From this foundation, the book moves to the critical question of reliability and resiliency. Modern grids must balance the intermittency of renewables with the demand for a continuous supply. Energy storage, smart grids, and innovative market mechanisms all play a role, but they also raise new questions about cost and feasibility. The trade-offs between reliability and affordability are particularly important as countries pursue net-zero emission targets.

Another central theme is the rise of electric vehicles and large-scale energy storage. These technologies promise to reduce dependence on fossil fuels, but they also reshape electricity demand and intensify the need for critical minerals. Their rapid adoption requires fresh thinking about infrastructure planning, recycling strategies, and grid integration.

The journey to net-zero is also tied to broader economic models. The book addresses the necessity of circular economy principles, where waste is minimized, materials are reused, and energy production is aligned with long-term ecological balance. This vision extends beyond renewables to include nuclear technologies, cross-border power grids, and even advanced concepts such as space-based energy transfer. By exploring both conventional and futuristic solutions, we aim to provide a complete picture of the options available.

A distinctive contribution of this work is the identification of twelve key research gaps. These gaps represent areas where knowledge, policy, or technology has not yet caught up with the urgency of the energy transition. They cut across generation, delivery, and consumption, and they are analyzed not only in technical terms but also in their social, economic, and environmental dimensions. The mapping of these gaps offers researchers and decision-makers a practical tool to prioritize action.

This book does not claim to offer all the answers. Instead, it provides a structured roadmap for asking the right questions and approaching the energy transition with clarity. Our hope is that it will serve as a bridge between academic research, policy development, and real-world application. By engaging engineers, economists, policymakers, and communities, we can collectively move toward systems that meet the world's growing demand without deepening environmental and social crises.

The vision we present is ambitious: a future where clean electricity is available to all, where affordability is no longer a barrier, and where reliability is ensured regardless of geography or economic status. Achieving this vision will require innovation, investment, and international cooperation. It will also require a willingness to challenge existing assumptions about energy systems and to embrace solutions that are both practical and forward-looking.

Ultimately, the goal of this book is to encourage dialogue, inspire research, and support informed decisions that bring us closer to universal access to sustainable, affordable, and reliable electricity. The path is complex and filled with trade-offs, but it is also rich with opportunity. By addressing the research gaps outlined here and exploring the strategies proposed, we can move toward an energy future that is fair, resilient, and sustainable for generations to come.

Wollongong, Australia

Md. Rabiul Islam
Behnaz Behi
Faisel Tubbal
Qinghua Guo
Sasha Nikolic
Raad Raad

Competing Interests The authors have no competing interests to declare that are relevant to the content of this manuscript.

Contents

About the Authors

Dr. Rabiul Islam received the Ph.D. degree from University of Technology Sydney (UTS), Sydney, Australia, in 2014 in electrical engineering.

He is currently an Associate Professor with the School of Electrical, Computer, and Telecommunications Engineering (SECTE), University of Wollongong (UOW), New South Wales, Australia. He is a Senior Member of IEEE. His research interests are in the fields of power electronic converters, renewable energy technologies, power quality, electrical machines, electric vehicles, and smart grid. He has authored or co-authored more than 400 papers, including more than 110 IEEE Transactions/IEEE Journal papers. He has written or edited nine technical books published by Springer and Taylor & Francis. He is serving as an Associate Editor for the IEEE Transactions on Industrial Electronics and IEEE Transactions on Energy Conversion. As a Lead Guest Editor, he has organized the first joint IEEE Industrial Electronics Society and IEEE Power and Energy Society Special Section entitled "Advances in High-Frequency Isolated Power Converters". He is an Editor of the Book Series entitled "Advanced in Power Electronic Converters" for CRC Press, Taylor & Francis Group. He is the Chair of IEEE New South Wales (Australia) Joint Chapter IAS/IES/PELS. He has organised many international conferences in different countries and was the Keynote Speaker for many international conferences. He has received several funding from Government and Industries including in total $5.48 million from Australian Government through Australian Research Council (ARC) Discovery Project (DP) 2020 entitled "A Next Generation Smart Solid-State Transformer for Power Grid Applications" and ARC Industrial Transformation Training Centre Project 2021 entitled "ARC Training Centre in Energy Technologies for Future Grids". He is one of the Chief Investigators of the ARC Training Centre in Energy Technologies for Future Grids and Theme Leader of Australian Power and Energy Research Institute.

Dr. Behnaz Behi received Ph.D. in Electrical Engineering from Murdoch University, Australia in 2022. Behnaz has developed an effective bidding strategy for realistic virtual power plants integrated with the wholesale electricity market consisting of solar PV, flow battery, heat pumps, and smart appliances. Behnaz has received a Science Industry Fellowship Award from the Australian Department of Jobs, Tourism, Science and Innovation, supported by the industry partner. Behnaz has been a research assistant at the University of Wollongong on developing a sustainable, reliable, and affordable model for future energy systems. At Murdoch University, Behnaz is working on planning zero emission renewable multi microgrids for a City Council in WA. Previously, Behnaz has completed several sustainable developments consultancy including economic, social, and environmental aspects in the usage of green natural and wind resources, water management, energy optimization in different sectors. Her interests include sustainable energy systems planning, renewable resource management, and energy optimization.

Dr. Faisel Tubbal achieved his educational milestones in the field of telecommunication engineering. He obtained his B.E. degree from the College of Electronic Technology in Tripoli, Libya, in 2004. He then pursued further studies and received a graduate diploma in engineering technology from the University of Wollongong in Wollongong, Australia, in 2011. Continuing his academic journey, he completed an M.S. degree in telecommunication engineering from the same university in 2012, followed by an M.S. degree in engineering management in 2013. Finally, in 2017, he successfully obtained his Ph.D. degree in telecommunication engineering, all from the University of Wollongong.

Throughout his career, Faisel has held various positions. He began his professional journey as a researcher at the Libyan Centre for Remote Sensing and Space Science (LCRSSC) in Tripoli, Libya. In 2012, he joined the University of Wollongong as an academic assistant at the School of Electrical, Computer, and Telecommunication Engineering. In recognition of his contributions, he was promoted to the role of labs manager at the School of Computer, Electrical and Telecommunication Engineering (SECTE) in 2019. Since 2017, Faisel has also served as a unit Convenor at the School of Computing, Engineering, and Mathematics at Western Sydney University, Australia. Presently, he holds the positions of Labs Manager, lecturer and Work Integrated Learning Coordinator at the School of Computer, Electrical and Telecommunication Engineering (SECTE) within the University of Wollongong.

As a researcher and author, Faisel has made significant contributions to the field. He has authored an edited book, three book chapters, and has published over 65 articles in esteemed conferences and journals. His research interests encompass various areas, including antenna designs for CubeSat applications, wearable antennas, antennas for medical applications, antenna designs using metamaterials, wireless power transfer using Rectenna technology and metasurface antennas.

Dr. Faisel Tubbal's dedication and expertise have been recognized through prestigious awards. He received the Vice-Chancellor's Awards for Outstanding Contribution to Teaching and Learning (OCTAL) in both 2016 and 2021. He is also honored as a Fellow at the Wollongong Academy for Tertiary Teaching and Learning Excellence (WATTLE). Additionally, he holds senior membership in the Institute of Electrical and Electronics Engineers (IEEE).

Dr. Qinghua Guo received the B.E. degree in electronic engineering and the M.E. degree in signal and information processing from Xidian University, in 2001 and 2004, respectively, and the Ph.D. degree in electronic engineering from the City University of Hong Kong in 2008. He is currently an Associate Professor with the School of Electrical, Computer and Telecommunications Engineering, University of Wollongong, Wollongong NSW, Australia, and an Adjunct Associate Professor with the School of Engineering, The University of Western Australia, Perth, WA, Australia. His research interests include signal processing, telecommunications, radar and optical sensing. He was a recipient of the Australian Research Council's inaugural Discovery Early Career Researcher Award in 2012.

Dr. Sasha Nikolic is a Senior Lecturer at the University of Wollongong and holds a Ph.D. in Engineering Education. He previously worked as a Laboratory Manager at the same university and spent several years in industry. Dr. Nikolic has received many teaching and learning awards, including an Australian Award for University Teaching citation in 2012 and 2019. He has a vibrant history of governance roles across the IEEE and is an executive committee member of the Australasian Association of Engineering Education. Sasha is also an Associate Editor for the *European Journal of Engineering Education* and the *Australasian Journal of Engineering Education.* He is also President of the Australasian Artificial Intelligence in Engineering Education Centre.

Dr. Raad Raad received a Bachelor of Engineering degree (Hons. 1) in Electrical Engineering from the University of Wollongong, Australia in 1997, and a Ph.D. degree entitled Neuro-Fuzzy Logic Admission Control in Cellular Mobile Networks, in 2006. Since 2004, he has been with the School of Electrical, Computer and Telecommunications Engineering, University of Wollongong and he is the current Head of School. His current research interests include wireless communications, CubeSat, the IoT, and Antenna design in addition to advanced power systems. He is part of the ARC ITTC for Future Grids and ARC hub for Connected Health Sensors, both major research initiatives.

Acronyms

AC	Alternating Current
BEV	Battery Electric Vehicle
CAPEX	Capital Expenditure
CCS	Carbon Capture and Storage
CE	Circular Economy
CO_2	Carbon Dioxide
COE	Cost of Electricity
DC	Direct Current
DER	Distributed Energy Resources
EU	European Union
EV	Electric Vehicle
GDP	Gross Domestic Product
GHG	Greenhouse Gas
HEV	Hybrid Electric Vehicle
HVDC	High Voltage Direct Current
IAEA	International Atomic Energy Agency
ICE	Internal Combustion Engine
ICT	Information and Communication Technology
IEA	International Energy Agency
IoT	Internet of Things
IPCC	Intergovernmental Panel on Climate Change
LCA	Life Cycle Assessment
LCOE	Levelized Cost of Energy
Li-ion	Lithium-ion
LWR	Light Water Reactor
MW	Megawatt
NPP	Nuclear Power Plant
NZE	Net-Zero Emissions

OECD	Organisation for Economic Co-operation and Development
OPEX	Operating Expenditure
PHEV	Plug-in Hybrid Electric Vehicle
PV	Photovoltaic
RE	Renewable Energy
RES	Renewable Energy Sources
RF	Radio Frequency
SDG	Sustainable Development Goal
SMR	Small Modular Reactor
UN	United Nations
VRE	Variable Renewable Energy
WHO	World Health Organization

List of Figures

1 Introduction to Sustainable, Affordable, and Reliable Electricity

Electricity lies at the heart of contemporary progress, shaping economic development, social welfare, and technological innovation. Yet the global pursuit of universal electricity access is challenged by rapidly growing demand, uneven distribution of resources, and the urgent need to reduce environmental impacts. Chapter 1 explores these interlinked dynamics by presenting the realities of rising consumption, the pressing threat of greenhouse gas emissions, and the difficulties faced by renewable energy technologies in meeting ambitious integration targets. It emphasizes that electricity affordability, reliability, and sustainability must be considered together, as none of these dimensions can be secured in isolation. By examining community-level challenges, global renewable energy strategies, and the motivations for transitioning to low-emission systems, the chapter lays the foundation for the critical discussions that follow throughout the book.

Energy Challenges Facing Communities

Modern civilization cannot exist without electricity since it is vital to economic growth, technical improvement, and human welfare in general. It fuels businesses, residences, medical facilities, educational institutions, and transit networks, promoting advancement and raising living standards for billions of people worldwide. The increasing urbanization and growth of the world's population is placing increasing pressure on water and energy supplies, resulting in an impending catastrophe with far-reaching consequences [1]. The increased consumption patterns promoted by modern lives, along with the growing need for energy to power homes, businesses, and transportation, increase the strain on finite fossil fuel supplies and hasten climate change. For instance, by 2040, the International

Md. R. Islam et al., *Universal Access to Sustainable, Affordable, and Reliable Electricity*, Synthesis Lectures on Sustainable Development,
https://doi.org/10.1007/978-3-032-19517-3_1

Energy Agency (IEA) predicts that the world's need for power will have increased by 62% from 2017 [1]. In a similar vein, the growing population places tremendous pressure on water supplies for industrial, agricultural, and drinking purposes, which causes resource depletion, contamination, and unequal distribution [2]. Industrialization's encouragement of increased consumption exacerbates these problems and puts additional stress on already overburdened resources. A multimodal approach is needed to address this dilemma, one that prioritizes equitable access to key resources for all. This includes investing in green and renewable energy sources, conserving water, encouraging sustainable consumption behaviors, and enacting regulations. Although investigating access to water and its quality is out of the scope of this book, access to quality energy and access to quality water are closely linked. For example, electricity powers water treatment plants that make water safe to drink and pumps that bring water to homes and businesses. If quick action is not taken, the problem might worsen and have serious consequences for both the present and future generations. This critical challenge is one of the *motivations* for developing this book to discover what are the gaps in establishing a sustainable approach in the use of resources and the usage of electricity.

The globe is currently dealing with serious issues, including the energy crisis, global warming, and emission reduction. Conventional techniques for producing power, including burning fossil fuels, have greatly increased greenhouse gas emissions, which have a negative impact on the environment and public health. Fossil fuel combustion releases greenhouse gases (GHGs), which have a wide range of negative effects on the environment and human health in addition to escalating global warming. Methane (CH_4), carbon dioxide (CO_2), and other greenhouse gases trap heat in the atmosphere, raising global temperatures and upsetting established climatic patterns. This phenomenon is responsible for the melting of ice caps, severe weather, and ecological disturbances [3]. Furthermore, burning fossil fuels releases pollutants into the air, soil, and water, including sulphur dioxide (SO_2), nitrogen oxides (NOx), and particulate matter. These pollutants can cause respiratory disorders, cardiovascular problems, and other health problems in people. Thus, it is clear from scientific research that immediate action is needed to make the shift to cleaner, more sustainable energy. Protecting the health of our world and its inhabitants requires actions to lessen these effects, such as switching to cleaner, renewable energy sources and adopting sustainable habits. This important challenge is another *motivation* behind writing this book to identify key research questions to be addressed in order to resolve GHG pollution associated with energy generation and consumption.

Renewable-based energy sources, including solar, wind, hydro, biomass, nuclear, and geothermal, have emerged as feasible alternatives that can meet the increasing electricity demand while mitigating the adverse effects of greenhouse gas emissions [4]. Renewable energy resources are widely regarded as low emission for sustainable energy production primarily due to their inherent characteristics and the processes involved in harnessing them. Unlike fossil fuels, which release significant amounts of greenhouse gases such as carbon dioxide when burned for energy production, renewable sources like solar, wind, hydro, and geothermal power generate electricity without emitting harmful pollutants or greenhouse gases. Solar energy relies on capturing sunlight through photovoltaic panels or concentrating solar power systems, while wind power harnesses the kinetic energy of moving air with turbines. Hydroelectric power utilizes the force of flowing water to generate electricity, and geothermal energy taps into the Earth's natural heat from within. These renewable sources offer a more sustainable alternative compared to fossil fuels, reducing carbon emissions, and mitigating the adverse impacts of climate change while promoting cleaner air and healthier environments. However, manufacturing and recycling renewable energy technologies are associated with GHG release and water and soil pollution. This *book is developed* to explore real emissions of technologies, including renewable energy ones, so decision-makers and investors can make informed decisions on low-emission technologies and methods.

The affordability of access to energy is impacted by the rising cost of energy and electricity, which is caused by a variety of factors such as changes in fuel prices, infrastructure investments, regulatory regulations, and technical improvements [5]. The increasing demand for energy worldwide, especially in developing countries, puts more strain on available energy supplies. This increases manufacturing costs, which in turn raises consumer prices. Globally, these growing expenses have a significant impact on people's development, health, and educational opportunities. Increased energy costs put pressure on household finances, leaving less money available for necessities like healthcare and education. Businesses also have to pay more for goods and services due to rising operating costs, which can worsen inflation and lower consumer purchasing power. Vulnerable populations may have to choose between paying for necessary medical care and keeping their houses warm enough, which could have a negative impact on their health. Moreover, higher energy costs can hinder educational opportunities, as schools may struggle to afford utilities, leading to inadequate facilities and resources for students. Ultimately, the increasing cost of energy and electricity poses significant challenges to socio-economic development and well-being, underscoring the urgent need for sustainable energy solutions and equitable access to affordable power. The falling costs of renewable energy technologies have made them increasingly competitive with conventional energy sources, opening new avenues for widespread adoption and increasing the affordability of electricity generation [6]. Therefore, the affordability of access to electricity is another challenge that *motivated* the development of this book to identify key research gaps towards affordable electricity delivery.

Despite significant advancements in energy infrastructure, a considerable portion of the global population still lacks access to electricity, primarily due to a combination of socio-economic, geographical, and political factors. In remote or rural areas, the cost of extending power grids can be prohibitively expensive, making it financially unfeasible for utilities to provide electricity to these communities [7]. In certain areas, political instability, conflict, and ineffective governance also obstruct the development of infrastructure and funding for energy access initiatives. Poverty also makes the issue worse because marginalized groups frequently lack the funds necessary to pay for recurring energy bills or to connect to the grid.

This book *discovers and discusses* a multifaceted approach to solve this urgent problem of access to affordable electricity. For example, to affordably supply electricity to rural populations, this involves making investments in off-grid renewable energy alternatives, including solar lamps, microgrids, and small-scale hydroelectric systems. Furthermore, enacting laws that are supportive of these initiatives, cultivating public-private partnerships, and enhancing governance can encourage investment in energy infrastructure and increase the number of marginalized people who have access to power. Furthermore, capacity-building and education programs that empower local people and promote energy efficiency are essential for sustainable energy access solutions. By addressing these challenges comprehensively, the global community can work towards ensuring universal access to electricity, driving socio-economic development, and improving the quality of life for millions worldwide.

Renewable Integration Targets

The global trend towards renewable integration is driven by the need to achieve sustainability and reduce greenhouse gas emissions [3]. Many countries are adopting a zero net emission strategy by 2050, however, as an example, Australia, Germany, and the USA are briefly reviewed here in terms of their zero net plans. 'Net zero emissions' refers to achieving an overall balance between greenhouse gas emissions produced and greenhouse gas emissions taken out of the atmosphere. These initiatives are significant as they aim to transition their energy systems to rely primarily on renewable sources such as wind, solar, and hydropower.

In Australia, the government has set a target of achieving net zero emissions by 2050. They are investing in renewable energy projects and implementing policies to support their transition towards clean energy sources [8].

Germany has been a pioneer in renewable energy integration. Their Energiewende (energy transition) policy aims to phase out nuclear power and coal while increasing the share of renewables in the energy mix. They have made significant progress in deploying wind and solar power, resulting in a substantial reduction in greenhouse gas emissions [9].

The USA has also made commitments to achieve zero net emissions by 2050. Several states and cities have implemented renewable energy targets and policies to promote the adoption of clean energy technologies. The federal government is investing in clean energy research and development and supporting initiatives to accelerate the deployment of renewable energy [10].

These initiatives have significant costs associated with them. The transition to renewable energy requires upfront investments in infrastructure development, including the construction of renewable power plants and the upgrading of transmission and distribution systems. In addition, there are considerable costs of operation and maintenance, and replacement costs associated with renewable energy technologies and energy storage. Renewable integration not only mitigates climate change but also promotes innovation, strengthens energy independence, and creates job opportunities in the clean energy sector. Therefore, one question is whether the long-term benefits of achieving sustainability and reducing greenhouse gas emissions outweigh the costs or not.

Australia has set an ambitious target of achieving zero net emissions by 2050, with a technology-based approach that prioritizes renewable energy and carbon capture and storage [11]. A $20 billion investment in low emissions technology is set to catalyze a remarkable $80 billion in total private and public investments in technologies such as clean hydrogen, carbon capture, and energy storage and innovations that reshape industries and communities.

Also, there is support to make ultra-low-cost solar a reality with the goal of solar electricity at an astonishing $15 per megawatt-hour (MWh). To reach this goal, Australia will need to make significant investments in renewable energy infrastructure and storage capabilities. According to the Net Zero Australia project, the country will require about 40 times the current total generation capacity of the national electricity market [12].

To meet the zero net emissions target, Australia will need an estimated 1,900 GW of solar power capacity and 174 GW of wind power capacity by 2050 [12]. These figures are substantial and highlight the scale of the plan. To put it into perspective, the required area for solar arrays alone would be nearly the size of Ireland [12].

The cost of achieving zero net emissions by 2050 is estimated to be around 8 trillion dollars [11]. This cost includes investments in renewable energy infrastructure, carbon capture and storage technologies, and other measures aimed at reducing emissions. It is important to note that such a significant investment could potentially lead to an increase in electricity prices.

The global pursuit of zero net emission targets presents significant challenges in achieving sustainability, affordability, and reliability of electricity supply for all. This endeavor is complex and requires careful consideration of various factors, including cost, infrastructure, technology, policy, and social acceptance.

Challenges Facing Renewable Technologies

Renewable energy sources offer promising solutions to mitigate climate change and transition towards a more sustainable energy future, but they also present significant challenges. One major obstacle is the intermittency of renewable sources such as solar and wind energy, which can fluctuate depending on weather conditions and time of day. The intermittent nature of solar and wind power, for example, presents technical and operational hurdles that maximize their contribution to a clean energy future. Overcoming these challenges requires innovative solutions and technological advancements that can effectively integrate renewable energy sources into the grid [13].

Policy frameworks play a crucial role in facilitating the transition towards renewable-based energy systems. Governments and regulatory bodies need to create supportive and enabling environments that incentivize investment, promote research and development, and establish clear targets for renewable energy deployment [13].

Effective policies can address market barriers, reduce investment risks, and provide long-term stability, attracting both public and private sector investments in renewable must be overcome to ensure a reliable and stable electricity supply [14].

Similarly, the geographical limitations of hydro and geothermal energy resources pose constraints on their widespread deployment. This intermittency necessitates the use of energy storage systems to store excess energy during periods of high production for use when renewable generation is low. Another challenge lies in the emissions associated with manufacturing the equipment required to capture and utilize renewable energy [15]. The production of solar photovoltaic panels, wind turbines, and energy storage systems involves the extraction and processing of raw materials, transportation, and manufacturing processes that emit greenhouse gases and other pollutants. Additionally, the disposal of end-of-life components poses environmental concerns. Addressing these challenges requires advancements in energy storage technologies, increased efficiency in manufacturing processes, and the adoption of sustainable practices throughout the lifecycle of renewable energy systems to minimize their environmental impact and maximize their contribution to a clean energy future. Overcoming these challenges requires innovative solutions and technological advancements that can effectively integrate renewable energy sources into the grid [13].

Policy frameworks play a crucial role in facilitating the transition towards renewable-based energy systems. Governments and regulatory bodies need to create supportive and enabling environments that incentivize investment, promote research and development, and establish clear targets for renewable energy deployment [13]. Effective policies can address market barriers, reduce investment risks, and provide long-term stability, attracting both public and private sector investments in renewable energy projects.

Motivations

Considering the above-mentioned challenges, universal access to electricity with three distinguished traits of being reliable, affordable, and sustainable is imperative for addressing global energy challenges and fostering inclusive socio-economic development. Ensuring that everyone has access to electricity means not only expanding energy infrastructure but also prioritizing the reliability and affordability of energy services while transitioning to sustainable sources. Renewable energy technologies offer a pathway to achieving this goal by providing clean and abundant energy resources that can be harnessed locally, reducing reliance on centralized power grids and fossil fuels. By investing in renewable energy projects and energy storage solutions, governments, businesses, and communities can enhance energy access while mitigating climate change and reducing harmful emissions associated with traditional energy sources. Moreover, promoting energy efficiency measures and fostering innovation in energy technology can further enhance the affordability and sustainability of electricity access for all, empowering individuals and communities to thrive in a more equitable and resilient energy future. Figure 1.1 shows the characteristics of the proposed electricity system in this book.

Therefore, the *motivation for this research* arises from the existing body of knowledge and the urgent need to accelerate the transition to reliable, affordable, and sustainable energy systems. This book aims to address critical research questions and fill gaps in the literature regarding the sustainable, affordable, and reliable supply, delivery, and consumption of electricity for everyone. By drawing insights from various aspects such as renewable energy integration, policy frameworks, and technological advancements, the book seeks to provide a comprehensive understanding of the challenges and opportunities in achieving sustainable electricity systems.

Contributions

Sustainability is a key aspect of achieving zero net emissions. The transition to renewable energy sources requires a massive scale of deployment. However, the upfront costs associated with building and maintaining renewable infrastructure can be substantial. This cost factor raises concerns about the affordability of renewable energy for all consumers, especially in developing countries, where access to affordable energy is crucial for economic growth and poverty alleviation. To this end, this book discusses that not all challenges associated with achieving sustainability have been identified and discussed. This book investigates the research gaps and topics that should be addressed to enable our communities to achieve their sustainability goals.

Affordability is another critical aspect that needs to be addressed, as emphasized by the World Bank. While renewable energy costs have been decreasing over time, the initial investment required for infrastructure development remains a challenge. This can lead to an increase in electricity prices, which may impact consumers' ability to afford reliable and sustainable energy. Moreover, the integration of intermittent renewable sources like solar and wind power requires additional investments in energy storage systems to ensure a reliable electricity supply. These storage technologies, such as batteries, can be costly and pose challenges to affordability. In this book, the roadblocks to cost-effective access to electricity are discussed, which shows that there are still many gaps between us and the goal of affordable electricity for all.

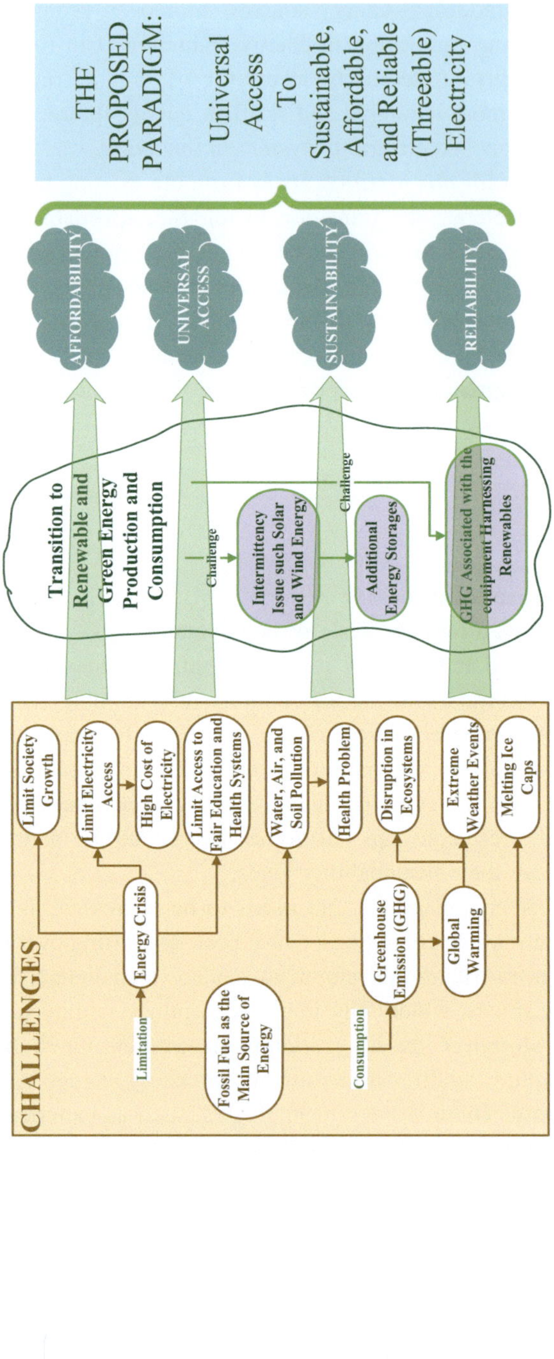

Fig. 1.1 The characteristics of the proposed electricity system in this book

Reliability of electricity supply is essential for the smooth functioning of societies and industries. Unlike conventional power plants, renewable sources are intermittent and dependent on weather conditions. This intermittency poses challenges to grid stability and requires investments in smart grid technologies, energy storage systems, and backup power sources. Balancing electricity supply and demand becomes more complex with a higher share of renewable energy in the grid. Ensuring a reliable and resilient electricity supply for all consumers, including those in remote areas, is a crucial challenge in achieving zero net emissions, as stated by the international energy agency (IEA). Therefore, achieving zero net emission targets globally while ensuring sustainability, affordability, and reliability of electricity supply for all is a complex task, as mentioned by intergovernmental panel on climate change (IPCC). The challenges encompass various factors, including cost, infrastructure, technology, policy, and social acceptance. In reliability of supply, there are some studies that should be carried out to develop a more reliable energy system in the presence of renewables at competitive prices. This book envisages these research gaps and research topics that enable reliability improvement for green energy systems.

This book aims to contribute to the understanding of the challenges associated with achieving sustainable, affordable, and reliable electricity supply worldwide. Specifically, it seeks to identify the most critical research questions and gaps that need to be addressed at different levels of electricity generation, transmission and distribution, and consumption. These challenges involve the development of innovative technologies for renewable energy generation, the establishment of robust infrastructure for efficient electricity transmission and distribution, and the promotion of sustainable consumption patterns.

By identifying these research questions and filling the discovered gaps, this book aims to provide policymakers, researchers, and industry stakeholders with a comprehensive understanding of the challenges and opportunities associated with achieving sustainable, affordable, and reliable electricity supply on a global scale. Ultimately, it seeks to contribute to the development of effective strategies and policies that can accelerate the transition to a low-carbon energy future while ensuring the well-being of communities and the planet.

This book identifies and discusses the challenging topic, presented in Fig. 1.2. These topics are associated with the challenges and research questions to be solved to achieve reliable, affordable, and sustainable electricity for everyone on our planet.

The main contributions of this book are as follows:

- Identify critical and main challenges facing communities and the electricity industry regarding energy access, availability, costs, and emissions,
- Identify twelve research gaps, including key questions to achieve universal access to sustainable, affordable, and reliable electricity,

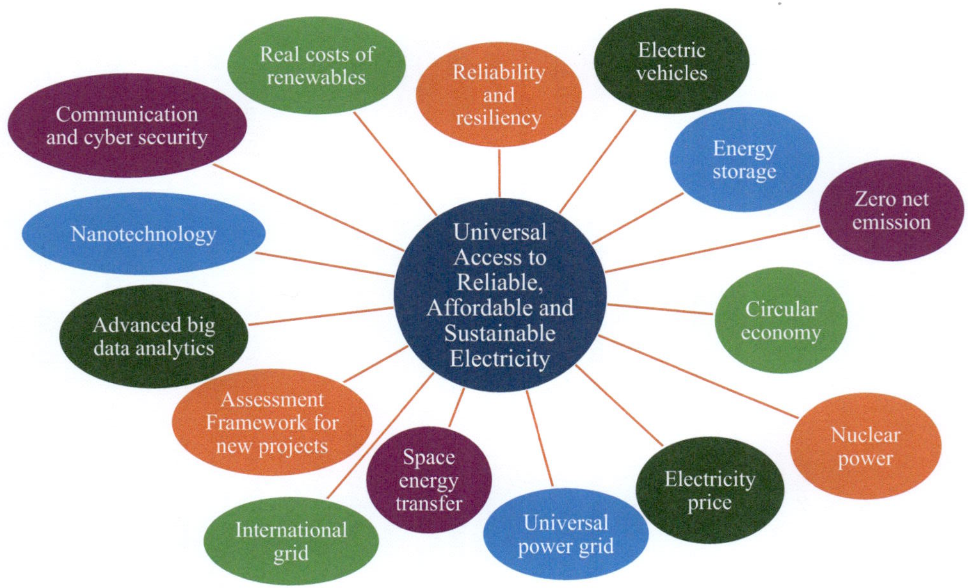

Fig. 1.2 The topics discussed in this book include the associated challenges and research questions aimed at achieving reliable, affordable, and sustainable electricity for everyone

- Develop critical research topics in each of the twelve research gaps, including the justification associated with technical, socio-economic, and environmental,
- Develop a mapping of research gaps into different sectors of the electricity industry, including generation, delivery, and consumption.

References

1. IEA, "World Energy Outlook 2019," IEA, Paris 2019, Available: https://www.iea.org/reports/world-energy-outlook-2019.
2. A. A. Ahmed, M. A. Nazzal, B. M. Darras, and I. M. Deiab, "Global warming potential, water footprint, and energy demand of shared autonomous electric vehicles incorporating circular economy practices," *Sustainable Production and Consumption,* vol. 36, pp. 449–462, 2023/03/01/ 2023.
3. I. M. Fund, "REACHING NET ZERO EMISSIONS," June 2021 2021, Available: https://www.imf.org/external/np/g20/pdf/2021/062221.pdf.
4. P. Denholm *et al.*, "The challenges of achieving a 100% renewable electricity system in the United States," *Joule,* vol. 5, no. 6, pp. 1331–1352, 2021/06/16/ 2021.
5. D. S. Mallapragada, C. Junge, C. Wang, H. Pfeifenberger, P. L. Joskow, and R. Schmalensee, "Electricity pricing challenges in future renewables-dominant power systems," *Energy Economics,* vol. 126, p. 106981, 2023/10/01/ 2023.
6. S. Algarni, V. Tirth, T. Alqahtani, S. Alshehery, and P. Kshirsagar, "Contribution of renewable energy sources to the environmental impacts and economic benefits for sustainable development," *Sustainable Energy Technologies and Assessments,* vol. 56, p. 103098, 2023/03/01/ 2023.
7. E. Young, "Electricity network services, Long-term trends in prices and costs," https://www.nsw.gov.au/sites/default/files/miscellaneous/electricity_network_services.pdf22 November 2013 2013.
8. (2022). *Australia joins international Net Zero Government Initiative.* Available: https://www.dcceew.gov.au/about/news/australia-joins-international-net-zero-government-initiative.
9. Y. Jafari, H. Engemann, T. Heckelei, and K. Hainsch, "National and Regional Economic Impacts of changes in Germany's electricity mix: A dynamic analysis through 2050," *Utilities Policy,* vol. 82, p. 101583, 2023/06/01/ 2023.
10. E. Mayfield, J. Jenkins, E. Larson, and C. Greig, "Labor pathways to achieve net-zero emissions in the United States by mid-century," *Energy Policy,* vol. 177, p. 113516, 2023/06/01/ 2023.
11. A. Government, "Australia's Long-Term Emissions Reduction Plan: A whole-of-economy Plan to achieve net zero emissions by 2050," 2021, Available: https://www.dcceew.gov.au/sites/default/files/documents/australias-long-term-emissions-reduction-plan.pdf.
12. Z. N. Australia, "Final modelling results," 2023, Available: https://www.netzeroaustralia.net.au/wp-content/uploads/2023/04/Net-Zero-Australia-final-results-full-results-pack-19-April-23.pdf.
13. K. Moustakas, M. Loizidou, M. Rehan, and A. S. Nizami, "A review of recent developments in renewable and sustainable energy systems: Key challenges and future perspective," *Renewable and Sustainable Energy Reviews,* vol. 119, p. 109418, 2020/03/01/ 2020.

14. X. Chen, K. C. Leung, and A. Y. S. Lam, "Power Output Smoothing for Renewable Energy System: Planning, Algorithms, and Analysis," *IEEE Systems Journal,* vol. 14, no. 1, pp. 1034-1045, 2020.
15. IEA. *Total mineral demand for clean energy technologies by scenario, 2010-2040.* Available: https://www.iea.org/data-and-statistics/charts/total-mineral-demand-for-clean-energy-technologies-by-scenario-2010-2040-2.

2 Reliability and Resiliency

The transition to renewable-rich power systems introduces profound challenges in maintaining the stability and dependability of the electricity supply [1]. This chapter addresses the concepts of reliability and resiliency as central pillars in designing modern grids, highlighting the tension between the variability of renewable resources and the need for continuous, high-quality power delivery. It examines how different levels of renewable penetration affect system adequacy, the economic implications of ensuring reliability, and the role of advanced storage technologies and demand-side flexibility in strengthening resilience. By analyzing both technical and cost dimensions, the chapter underscores the importance of developing strategies that safeguard reliability while supporting the broader goal of sustainable energy transformation.

The Impact of a High Level of Renewable Integration on Reliability and Its Costs

In general, if additional renewable-based generations are added to the power system, the reliability of the system will increase [2]. However, if conventional generators are to be replaced by renewables, the intermittent nature of renewable energy sources poses challenges to power systems' reliable and resilient operation because renewables cannot provide electricity at all times [3]. Therefore, we need to put in more renewables and storage. Integrating more storage and renewables can enhance system reliability and resiliency. Simulations show that the expected energy not supplied (EENS) can be reduced by 30% by adding 70 MWh of battery storage and properly managing it for a 33-bus distribution network [4]. This EENS reduction or reliability improvement is then saturated

Md. R. Islam et al., *Universal Access to Sustainable, Affordable, and Reliable Electricity*, Synthesis Lectures on Sustainable Development,
https://doi.org/10.1007/978-3-032-19517-3_2

by increasing more batteries in the network [4]. The study demonstrates the cost reduction and enhanced reliability benefits of ESS charging and discharging in microgrids. This research identifies that weekly scheduling improves aging costs and reliability, whereas daily scheduling optimizes overall expenses [4].

A long-term reliability analysis in France's power grid by 2050, incorporating short-term operational conditions, is also conducted to see the impact of zero net emission programs [5]. This research combines system adequacy and transient stability considerations to quantitatively assess the reliability of the French power sector, defining a reliability indicator linked to kinetic reserves. The findings reveal that grid reliability can be maintained only for accommodating about 63% of renewable energy sources in the installed capacity without compromising system reliability, as shown in Fig. 2.1 [5]. However, it comes at the cost of increased system complexity and higher investment requirements, which ultimately lead to higher electricity prices. In the 100% renewable energy (100EnR) scenario for France, there is a significant increase in storage technology capacity, reaching 1.3 GW in 2050, a multiplication of approximately eight compared to the business as usual (BAU) scenario [5].

The presence of a reliability constraint leads to the installation of even more storage devices, with a multiplication of approximately six in the 100EnR scenario [5]. Also, the cost associated with this increase is significant, for example, for France, it needs an additional 100 billion Euro cash flow in 2050 to run the system for 100% renewable at the required reliability level, which is almost triple if we consider the BAU scenario [5]. As seen, reliability is an important factor in the final cost of energy systems and their secure operation, which should be considered in zero net emission planning in all regions [6]. Another study for Bali, Indonesia, shows that if the reliability cost is higher, the share of

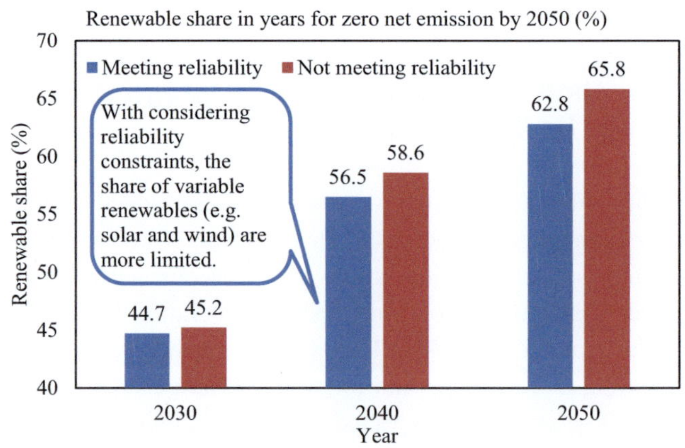

Fig. 2.1 Renewable share in 2030, 2040, and 2050 for zero net emission by 2050 (%)

renewable integration should be lower to guarantee the required level of reliability using dispatchable generations such as gas turbines or coal-fired power plants [7].

The impact of the required reliability level on the economic factors of electrification is also studied in [8]. Notably, the research reveals that overall costs tend to rise as expected reliability increases, with costs surging over sixfold when expected adequacy jumps from 10% to nearly 100%. This underscores the substantial cost implications associated with high-reliability expectations with renewables. Interestingly, optimizing energy portfolios alone may not effectively control overall costs. Instead, it suggests that incorporating technologies like demand response and system flexibility holds promise in simultaneously enhancing reliability and constraining overall costs within microgrids [8].

Furthermore, different levels of reliability in several pathways for achieving zero net emissions for the USA grid are studied [9], as seen in Fig. 2.2 in which selected results are illustrated. As seen, the cost of a zero net emission energy system with 100% renewables and higher reliability is 53% higher than the zero net emission system without higher reliability. Also, the levelized cost of energy (LCOE) is 32% higher for a zero net emission energy system with 100% renewables and higher reliability compared to the LCOE of the usual zero net emission system [9]. LCOE is indicative of how much customers will pay for energy, and it can be seen that both the level of reliability and the level of renewable are directly impacting this amount. When considering the least cost approach for a zero net energy system without any specific requirement for renewable share, the proposed renewable share would be around 57% by 2050 for the USA [9].

In addition, the cost of a zero net emission system with no renewable share requirement with a higher reliability increases by 13%, compared to the system without higher

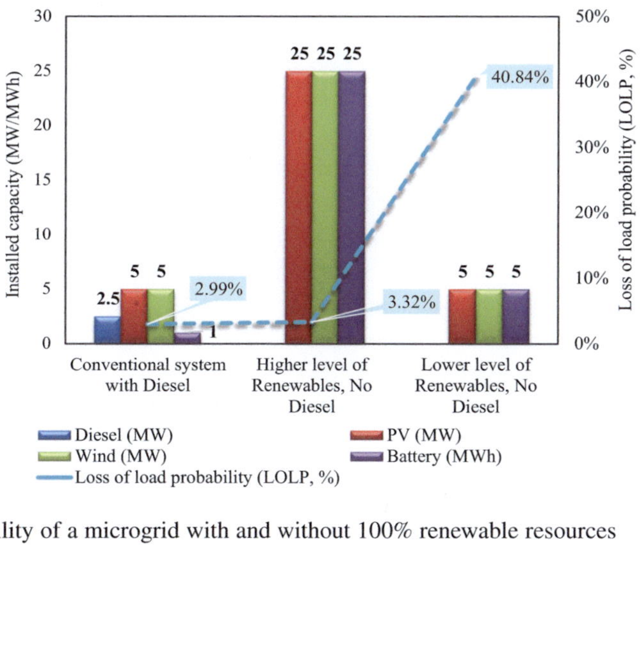

Fig. 2.2 Reliability of a microgrid with and without 100% renewable resources

reliability [9]. The main source of higher reliability that is examined in this study has been the flexibility available from loads, for which it's assumed that 17% of total demand is available to the grid to provide the required services for improving reliability [9]. Another analysis [10] investigates the reliability of microgrid systems, addressing the failure rates of equipment reliability and the variability of renewable energy sources simultaneously. It is important to consider the failure of equipment such as inverters in the analysis of reliability. For example, the mean time between failure of a PV inverter can be reduced from 12 years to 1.6 years after 10 years of usage [10]. Also, control of inverters and heat management can impact the reliability and lifetime of power electronic modules [11].

Reliability Improvement at Minimum Cost

Considering the significant impact of the high level of renewable integration on reliability, improving reliability to the required level at minimum cost is a critical mission in zero net planning. To improve reliability, various methods were examined in [12, 13], such as enhancing individual component reliability, increasing battery size, and expanding electricity generation capacity (both renewable and diesel). The results of these studies show that the reliability of microgrids reduces significantly if the dispatchable generator, such as a diesel generator removed from the grid [12, 13].

Also, it shows that to achieve a similar reliability level as the conventional resource, we need to install much more battery and renewable resources [12, 13]. For example, the LOLP of microgrids with a diesel generator and PV, wind and storage is about 3%, while without a diesel generator, the LOLP increases significantly to about 40%, which shows a dramatic fall in the reliability of the microgrid. To achieve a similar level of reliability for a 100% renewable-based microgrid as the system with diesel, we need to install 5 times more PV, wind, and battery, as shown in Fig. 2.3 [12].

In highly renewable-based power systems, recent research [14] has focused on establishing cooperative market mechanisms where renewable generators are treated as negative uncontrollable loads and can be mitigated within an aggregator [14]. However, the cooperative approach limits overall production and economic efficiency for both society and renewable generators. The primary challenge lies in defining positive marginal costs for renewables. To address this issue, the study suggests considering reliability as the production of renewable generators and introduces a novel non-cooperative reliability competition mechanism for renewable market participation. This non-cooperative approach proves more economically efficient, providing clearer signals for generation reliability costs, and ultimately enables the high renewable-based power system [14].

In order to quantify reliable service with renewables, the marginal cost of reliability improvement in the context of renewable-based microgrids should be determined over the long term, such as a 25-year lifecycle [15]. This parameter is a valuable metric for comparing the cost-effectiveness of renewable microgrids, aiding in decisions regarding

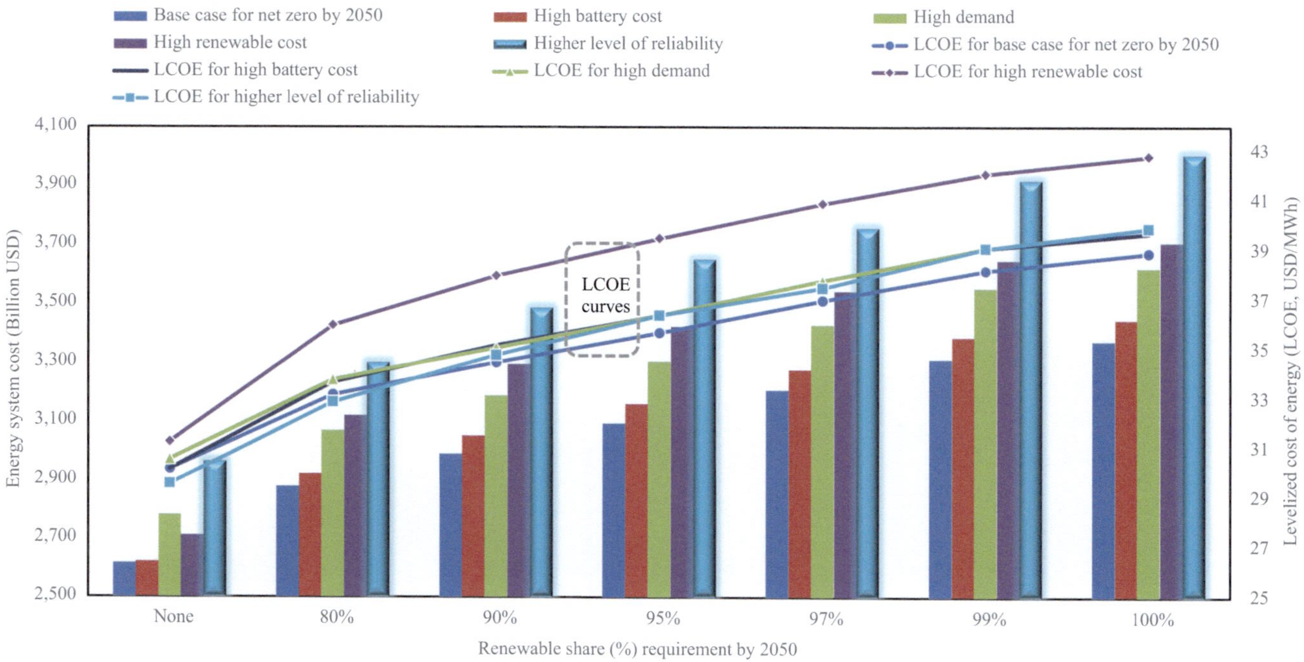

Fig. 2.3 Energy system cost and levelized cost of energy (LCOE) for a zero net emission target by 2050 at different requirements for the renewable share, including cases with higher reliability

reliability levels and imports/exports between microgrids. It also addresses changes in reliability requirements from customers and energy traders. A reliability contract between renewable resources and dispatchable resources, such as gas power plants, can also be useful, in which gas power plants can cover the load when renewables cannot provide the full amount of committed load in a market [16, 17]. This idea will highlight the concept of a multi-energy system in which reliability can be handled by certain resources to increase renewable integration. For example, the potential of hydrogen utilization in the transportation sector is on the rise [18]. To harness this potential and increase renewable energy integration in power systems, surplus electricity can be employed for water electrolysis, generating hydrogen. This approach facilitates the efficient coupling and synergy between various energy sectors, including electricity, heat, and gas, to enhance reliability and renewable integration [18]. Detailed research should be conducted for reliability as a service and the marginal cost of reliability in a market for renewable generators, both in terms of definition and market mechanism.

The Impact of Energy Storage on Reliability

As discussed, energy storage can play an important role in improving reliability in renewable-rich power grids. The strategy of charging and discharging energy storage impacts the level of reliability obtained by energy storage [19]. Strategies such as smoothing renewable energy fluctuation, maximizing revenue, and coordinating between maximum revenue and minimum renewable energy fluctuation can be established for energy storage. The strategy needs to be selected considering the network situation and the electricity market, however, in general, a maximizing revenue strategy can be useful for improving reliability [19].

Furthermore, energy storage can manage uncertainties associated with renewable generations, that are another critical factor in the reliability level of generation. This significant challenge of addressing uncertainties in generation availability from renewable sources can be tackled by optimal battery reconfiguration and restoration of grids to enhance customer-side reliability and overall system performance in distribution systems [20]. It emphasizes the importance of research on accounting for uncertainties related to power output from renewable resources, time-varying load demand such as residential and commercial load profiles, stochastic prediction errors, and random fault events in the evaluation of reliability and the development of optimal restoration strategies [20].

Energy storage can also provide inertia to power systems. System inertia is another critical parameter of power systems to determine grids' stability and resiliency [21]. Exploring various future scenarios for power grids with increasing levels of renewable energy generation can help anticipate declines in system inertia, which could potentially trigger grid instability. Some researchers [21] have leveraged historical data to assess the impact of variable renewable energy generation on system inertia, a key indicator of grid stability defined as the grid's ability to withstand contingency events, such as outages, without dropping below its standard margins of operating frequency. Rotational inertia, representing the kinetic energy in a generator's spinning rotor, contributes to system inertia, which encompasses the total available rotational inertia from synchronous generators. Studies indicate that grids with reduced system inertia levels, such as grids with high levels of renewable generation, require quicker frequency control following contingency events, as delayed responses can lead to issues like generator damage, load shedding due to under-frequency conditions, and, in the worst-case scenario, fault cascades and blackouts [21].

The impact of different energy storage technologies should also be studied as they influence the affordability and technical parameters of energy systems. Some energy storage, such as compressed air or gravitational, can provide energy for a longer period, which enhances the reliability of the system. Some other energy storage, such as lithium-ion, can react fast to events in power systems, so they contribute to the stability improvement of the grid. Therefore, deep research should be conducted to find out the optimal configuration of energy storage for the long-term planning of power systems, especially with the zero net emission target. Different energy storage technologies are discussed in the "Energy Storage" section in this book.

In such planning, fast reliability analysis methods should be utilized to avoid the long processing time of power system planning, considering the schedule of batteries at different levels of loads and renewables. For example, the application of the proposed reliability evaluation technique in [22] for large distribution planning, as illustrated through a 747-node network, substantially reduces computational time. Traditional methods took an extended 61 days to complete the task, whereas the new approach accomplishes it in a mere 4.6 hours, showcasing significant time efficiency improvements [22].

Balancing the need for increased renewable penetration while maintaining affordable electricity prices is still a critical challenge for power system planners and operators. Therefore, exploring the trade-offs between reliability, resiliency, and the cost of renewable-rich power systems is essential. It is required to examine various approaches to mitigate the intermittency challenge and optimize system performance.

References

1. P. S. Meera and S. Hemamalini, "Reliability assessment and enhancement of distribution networks integrated with renewable distributed generators: A review," *Sustainable Energy Technologies and Assessments,* vol. 54, p. 102812, 2022/12/01/ 2022.
2. I. Akhtar, S. Kirmani, and M. Jameel, "Reliability Assessment of Power System Considering the Impact of Renewable Energy Sources Integration Into Grid With Advanced Intelligent Strategies," *IEEE Access,* vol. 9, pp. 32485-32497, 2021.

3. W. Li, F. Dong, Z. Ji, and L. Ji, "Evaluation of provincial power supply reliability with high penetration of renewable energy based on combination weighting of game theory-TOPSIS method," *Sustainable Energy, Grids and Networks,* vol. 35, p. 101092, 2023/09/01/ 2023.
4. E. Rahmani, S. Mohammadi, M. Zadehbagheri, and M. Kiani, "Probabilistic reliability management of energy storage systems in connected/islanding microgrids with renewable energy," *Electric Power Systems Research,* vol. 214, p. 108891, 2023/01/01/ 2023.
5. G. S. Seck, V. Krakowski, E. Assoumou, N. Maïzi, and V. Mazauric, "Embedding power system's reliability within a long-term Energy System Optimization Model: Linking high renewable energy integration and future grid stability for France by 2050," *Applied Energy,* vol. 257, p. 114037, 2020/01/01/ 2020.
6. J. Beyza and J. M. Yusta, "The effects of the high penetration of renewable energies on the reliability and vulnerability of interconnected electric power systems," *Reliability Engineering & System Safety,* vol. 215, p. 107881, 2021/11/01/ 2021.
7. Y. Tanoto, N. Haghdadi, A. Bruce, and I. MacGill, "Reliability-cost trade-offs for electricity industry planning with high variable renewable energy penetrations in emerging economies: A case study of Indonesia's Java-Bali grid," *Energy,* vol. 227, p. 120474, 2021/07/15/ 2021.
8. J. Luo, H. Li, G. Huang, and S. Wang, "A multi-dimensional performance assessment framework for microgrids concerning renewable penetration, reliability, and economics," *Journal of Building Engineering,* vol. 63, p. 105508, 2023/01/01/ 2023.
9. W. J. Cole *et al.*, "Quantifying the challenge of reaching a 100% renewable energy power system for the United States," *Joule,* vol. 5, no. 7, pp. 1732–1748, 2021/07/21/ 2021.
10. D. M. Bui, P. D. Le, T. M. Cao, H. T. Nguyen, T. V. Tran, and H. M. Nguyen, "Boost-converter reliability assessment for renewable-energy generation systems in a low-voltage DC microgrid," *Energy Reports,* vol. 8, pp. 821–835, 2022/12/01/ 2022.
11. M. M. Islam and M. R. Islam, "Power loss and thermal modeling of charger circuit for reliability enhancement of EV charging systems," in Transportation, Cable Based and Wireless Charging Systems for Electric Vehicles: Technology and control, management and grid integration: Institution of Engineering and Technology, 2021, pp. 289–325. [Online]. Available: https://doi.org/10.1049/pbtr032e_ch11.
12. S. Nallainathan, A. Arefi, C. Lund, A. Mehrizi-Sani, and D. Stephens, "Reliability Evaluation of Renewable-Rich Microgrids Using Monte Carlo Simulation Considering Resource and Equipment Availability," in *2020 IEEE International Conference on Power Systems Technology (POWERCON),* 2020, pp. 1–6.
13. M. Firouzi, A. Samimi, and A. Salami, "Reliability evaluation of a composite power system in the presence of renewable generations," *Reliability Engineering & System Safety,* vol. 222, p. 108396, 2022/06/01/ 2022.
14. X. Ruhang and J. Jia, "Towards reliability competition: Non-cooperative market mechanism under high variable renewable energy penetration," *Applied Energy,* vol. 331, p. 120415, 2023/02/01/ 2023.
15. N. Sakthivelnathan, A. Arefi, C. Lund, A. Mehrizi-Sani, and D. Stephens, "Marginal Cost of Reliability Improvement for Standalone Microgrids," in *2021 31st Australasian Universities Power Engineering Conference (AUPEC),* 2021, pp. 1–6.
16. D. D'Achiardi, N. Aguiar, S. Baros, V. Gupta, and A. M. Annaswamy, "Reliability Contracts Between Renewable and Natural Gas Power Producers," *IEEE Transactions on Control of Network Systems,* vol. 6, no. 3, pp. 1075-1085, 2019.
17. Y. Wang, Y. Yang, and Q. Xu, "Reliability Assessment for Integrated Power-gas Systems Considering Renewable Energy Uncertainty and Cascading Effects," *CSEE Journal of Power and Energy Systems,* vol. 9, no. 3, pp. 1214-1226, 2023.

18. M. Zhang *et al.*, "Optimal Design and Operation of Regional Multi-Energy Systems With High Renewable Penetration Considering Reliability Constraints," *IEEE Access,* vol. 8, pp. 205307-205315, 2020.
19. J. Zhang, J. Liu, L. Chen, L. Zhang, P. Zeng, and Y. Li, "Reliability evaluation of high permeability renewable energy distribution network considering energy storage charge and discharge strategy," *Energy Reports,* vol. 9, pp. 361–368, 2023/05/01/ 2023.
20. K. Zou, G. Mohy-ud-din, A. P. Agalgaonkar, K. M. Muttaqi, and S. Perera, "Distribution System Restoration With Renewable Resources for Reliability Improvement Under System Uncertainties," *IEEE Transactions on Industrial Electronics,* vol. 67, no. 10, pp. 8438-8449, 2020.
21. S. C. Johnson, D. J. Papageorgiou, D. S. Mallapragada, T. A. Deetjen, J. D. Rhodes, and M. E. Webber, "Evaluating rotational inertia as a component of grid reliability with high penetrations of variable renewable energy," *Energy,* vol. 180, pp. 258–271, 2019/08/01/ 2019.
22. A. Arefi, G. Ledwich, G. Nourbakhsh, and B. Behi, "A Fast Adequacy Analysis for Radial Distribution Networks Considering Reconfiguration and DGs," *IEEE Transactions on Smart Grid,* vol. 11, no. 5, pp. 3896-3909, 2020.

3 Electric Vehicles and Energy Storage

The rapid growth of Renewable Energy Sources (RES) within the electric grid is driven by reduced reliance on traditional energy sources and increased demand for power to support economic growth. As discussed, various industries, including manufacturing, transportation, and communication, have witnessed a significant surge in their demand for electrical energy. RES play a pivotal role in meeting this growing energy demand, as they are environmentally friendly, scalable, and applicable across all sectors. However, RES have limitations, such as non-dispatchability and intermittency due to the variability of renewable resources such as wind and solar. To address such uncertainties, energy storage systems can be used, so developing and deploying appropriate energy storage is crucial. In addition, the rise of electric vehicles (EVs) promises a revolution in transportation. However, their proliferation comes with its own set of challenges. As EVs become more prevalent, the demand for power surges significantly. Some studies investigated the impact of EVs on electricity peak demand and energy consumption.

Impact of EVs on Electric Grids

A team of researchers from the University of Queensland conducted an original investigation [1] involving the analysis of detailed driving and charging data from 239 Tesla vehicles in Australia. This dataset covered a span of six months. It was observed that the average annual energy consumption for each EV amounted to approximately 2,143 kWh. Also, the data revealed distinctive patterns, including energy consumption peaks during the middle of the night and day, along with notably lower charging activity during peak

Md. R. Islam et al., *Universal Access to Sustainable, Affordable, and Reliable Electricity*, Synthesis Lectures on Sustainable Development,
https://doi.org/10.1007/978-3-032-19517-3_3

demand periods. Specifically, the per-EV impact during peak hours ranged from 150 to 300 W, with the lowest point occurring at 8 pm.

A study investigated the impact of plug-in hybrid electric vehicles (PHEVs) with different charging levels, namely 1.44 kW, 3.88 kW, and 7.68 kW, on electricity demand in the United States [2]. This study shows that charging at high power levels, whether at home, work, or any location, leads to a notable increase in peak power demand, ranging from about 50 to 62%. This heightened demand results from the reduced charging time for EVs, potentially enhancing the all-electric vehicle miles travelled, but increasing the daily energy drawn from the grid. Specifically, high charge powers contribute to a 6.6 to 11% increase in daily electric peak demand. In cases where only home and work charging are allowed, a lower power level of 1.44 kW, commonly found in households, helps distribute the evening load more evenly during off-peak nighttime hours. In contrast, higher power levels such as 3.88 and 7.68 kW introduce greater load variability and increase power demand during peak periods, particularly in the early morning and early evening.

Origin Energy in Australia, with the support of ARENA funding, conducted a trial involving 150 EV drivers to assess their charging behavior in various scenarios [3]. In the initial baseline condition, where no incentives were offered for EV charging time, approximately 30% of energy consumption by the vehicles occurred between 3 and 9 pm, as a peak period for the grid. This resulted in an average load per EV during the six-hour peak demand period of around 400 W, even though incentives to avoid peak-time charging were deliberately eliminated.

The first experiment introduced a price-based incentive, where users received a 10 cents per kWh benefit for avoiding EV charging during the 3 to 9 pm window. This incentive significantly reduced energy consumption during this peak period to just 10% of the total, resulting in a peak demand impact per EV of approximately 167 W. The second experiment involved external coordination through smart charging equipment, coupled with a daily 25-cent payment to incentivize consumers to accept external control, in addition to the 10 cents per kWh benefit from the first experiment. Under this condition, 6% of EV energy consumption took place during the 3 to 9 pm window, resulting in a peak demand impact per EV of about 100W.

These experiments show that financial incentives will reduce EV charging during peak by 20%, as shown in Fig. 3.1.

Considering the state government's targets of achieving 50% of new vehicle sales as EVs by 2030, there will be 1.5 million EVs on the road by 2030. Accordingly, the peak demand increase associated with this level of EVs would be around 600 MW. This increase accounts for approximately 2% of the total combined peak demand across the national electricity market (NEM) and the Southwest Interconnected System (SWIS). Although the peak demand increase is not significant, however, the behaviour and the usage pattern change of EV drivers should be studied to have a better understanding of the impact on the peak demand.

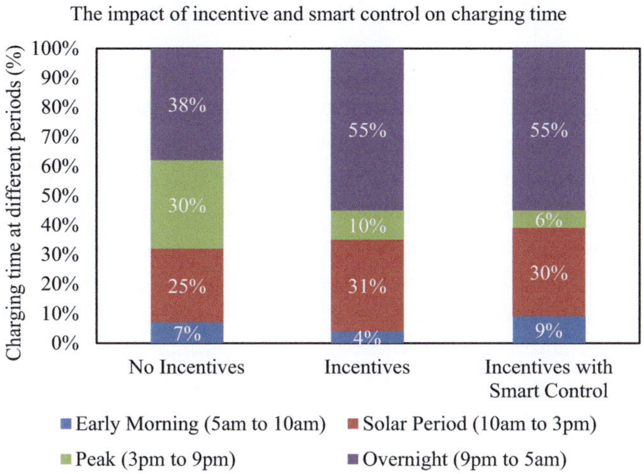

Fig. 3.1 The impact of incentive and smart control on charging time-based on the origin energy's experiment

The total energy consumption by EVs in 2023, considering the above assumptions, reaches about 3.22 TWh, which is around 1.3% of total electricity or 5.4% of total residential electricity in Australia in 2030. Although this amount is not significant, assuming 1.5 million EVs in Australia in 2030, these EVs are not distributed evenly around the country, therefore, in some suburbs might be severe congestion due to the high number of EVs.

Another important study needs to be conducted to examine the impact of ultra-fast charging of EVs on distribution grids [4]. These ultra-fast chargers have the ability to charge EVs in 5 to 10 minutes, however, they need a much higher charging capacity up to 350 kW (compared to the fast chargers with 50 kW capacity), which increases the tension on the network. The stability of electric grids needs to be studied considering such ultra-fast chargers. In order to stabilize grids due to the presence of EVs, we need to invest in additional equipment or services [5], therefore, the associated expenses should be considered when studying the cost and benefit of EVs.

Due to the above-mentioned impacts of EVs on electric grids, techniques such as reconfiguration of grids, dispatching distributed generations, and coordinating EVs [6, 7] to mitigate unbalance, improve voltage quality, and reduce losses in distribution grids. The role of superconducting cables and solid-state transformers for adding flexibility and capacity to electric grids is also critical [8]. Also, the use of local renewable resources and energy storage should be investigated to explore the possibilities of reducing the impact of EV charging on grids [9]. While incentivizing EV owners to move away from the peak demand is effective, however, this should be studied very clearly how much incentive is optimum. To this aim, a joint planning of grids and EV charger locations, and incentives is required [10–12].

All the above-mentioned grid planning and control of EVs create inconveniences for EV owners and necessitating a complex communication infrastructure and substantial data processing. An innovative strategy to address this issue would be a centralized control system built upon Internet of Things (IoT) technology and leveraging an artificial intelligence (AI) optimization algorithm to coordinate the distribution of both EVs and distributed energy resources (DES) [13]. On the other hand, it is critical to consider the cost associated with this infrastructure when considering the affordability and integration of EVs within future grids. The cyber and physical security of such a communication and control system is also very critical [14] to ensure proper and reliable operation of the grid, EVs, energy resources, and energy storage.

Emissions and Minerals of EVs

Transport is the third largest source of emissions in Australia. By promoting EVs, the government has a plan to cut emissions by at least 3 million tonnes of carbon by 2030, and over 10 million tonnes by 2035. The government has already cut taxes on EVs through the Electric Car Discount, saving up to $11,000 a year on a $50,000 electric vehicle.

To comprehensively compare the overall life-cycle greenhouse gas (GHG) emissions of various passenger cars with diverse powertrains and energy sources, it is required to quantify the total life-cycle GHG emissions in production, utilization, and recycling emissions [15]. A study investigated these emissions for a total of 790 different vehicle variants [15]. This study shows that conventional gasoline and diesel vehicles emerge as the largest contributors to total life cycle GHGs when compared to vehicles utilizing alternative energy sources. When powered by green electricity, plug-in hybrid electric and fully electric vehicles can reduce life-cycle emissions in comparison to combustion engine vehicles by 73 and 89%, respectively. Similarly, biogas-powered vehicles achieve an 81% reduction in emissions. Fuel cell vehicles, using commercially available grey hydrogen, achieve a comparable GHG emission reduction of 60% when compared to electric vehicles charged with conventional electricity [15].

However, it should be noted that, as per the report by IEA, each EV needs 7 times more minerals than each conventional car [16]. The mining, extraction, and production of such minerals will result in land and water pollution and damage, in addition to the high amount of energy consumed for the processing of such minerals. Therefore, a comprehensive study needs to be conducted to include the impact of mining minerals in the production of EVs to calculate the accurate figure of the impact of EVs on emissions and the sustainability of land and water.

When promoting EVs, it is essential to supply them with green or low-carbon electricity, otherwise, the total GHG of EVs will be high, as shown in Fig. 3.2. As seen, if the

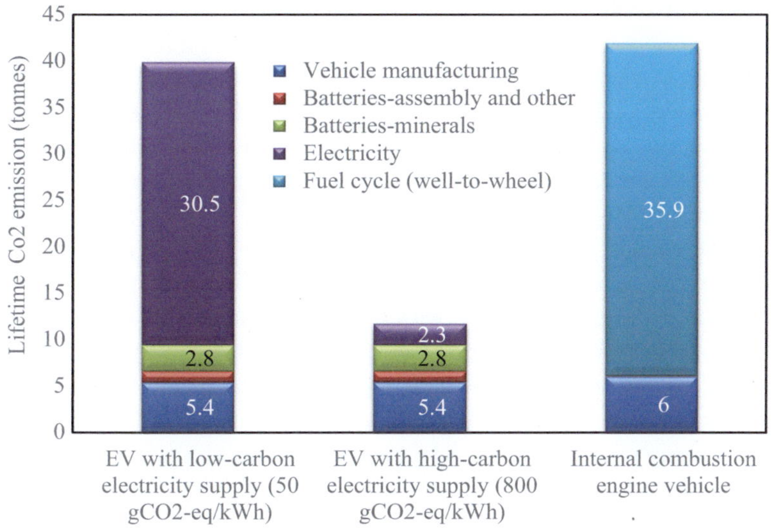

Fig. 3.2 Tonnes of CO_2 equivalent per vehicle lifetime in different categories of manufacturing and operation

electricity that is supplied to EVs has a high level of emission, e.g. 800 gCo_2-eq/kWh, the total emission of an EV and a conventional car are almost the same. For example, according to AEMO [17], the equivalent CO_2 emission per kWh for the NEM area is about 620 g in 2023 so far, which is high enough to reduce the impact of EVs on emission reduction.

Autonomous EVs in a Shared Economy

The increasing number of EVs necessitates a risk management approach to ensure grid security and minimizing life-cycle GHG emissions [18]. The new concept of autonomous EVs' shared economy can provide risk management potential for the power system reliability which needs to be explored. A shared economy means that people share EVs [19] or charging points [20] Among themselves, not individually, those assets are owned. Autonomous EVs refer to EVs with the capability of self-driving or semi-driving.

The autonomous EV shared economy presents a promising solution for managing the amount of minerals used for EVs and EVs' power consumption. By leveraging shared EV ownership and incorporating self-driving or semi-driving capabilities, such risk can be minimized. The sharing concept will also reduce the number of required EVs.

Energy Storage Technologies

The effective use of energy storage systems (ESS) can address the shortcomings of varying renewables, with options for storing energy generated by RES and giving energy back to the users and grid when they need to reduce the impact of intermittency [21]. Different types of energy storage can be used towards maximizing the integration of renewables [22, 23]. Energy storages are categorized in mechanical (such as compressed air and pumped hydro), thermal (such as cryogenic energy storage), chemical (such as Li-ion and Ni-Cd), and electrical (such as superconducting magnet energy storage). Figure 3.3 shows the characteristics of some energy storage with the scaled parameters for the comparison.

As seen, the most affordable options are pumped hydro energy storage (PHES) and compressed air energy storage (CAES), as their lifetime is the highest, but $/kW is low with a very high-power rating. However, the response speed of PHES and CAES is less than chemical energy storage, such as Li-ion and Ni-Cd. Also, chemical energy storage is easily scalable and can contribute to the stability of power systems. In most cases, a combination of different types of energy storage is the best option in terms of cost and power quality and reliability of grids [24]. As costs continue to decrease, energy storage is assuming a broader role in energy markets, transitioning from niche applications like grid balancing to more extensive functions such as substituting conventional power generators for reliability, operating reserves, ancillary services for grid stability, delivering power quality services, and facilitating renewables integration. Energy storages also play

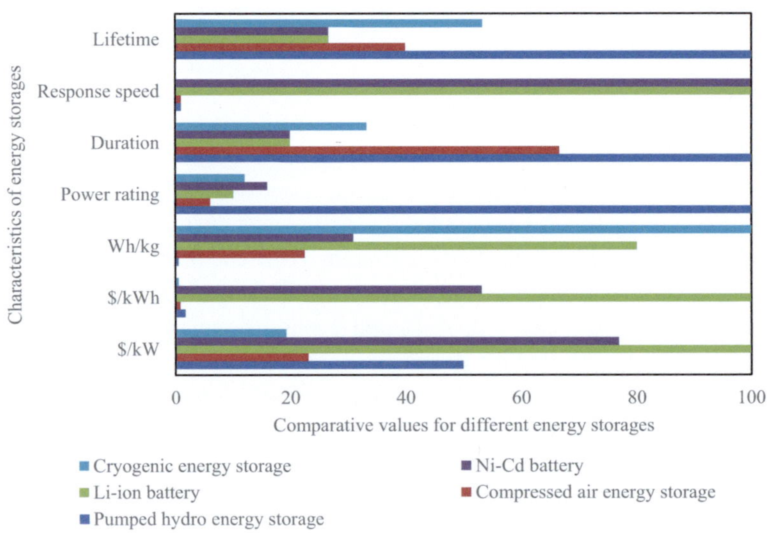

Fig. 3.3 Characteristics of different types of energy storage

a pivotal role in off-grid systems, regulating frequency, managing power fluctuations, and ensuring stability.

As per the International Energy Agency (IEA), Pumped-storage hydropower stands out as the predominant storage technology in widespread use, offering substantial untapped potential in numerous geographic areas. Among grid-scale storage solutions, batteries emerge as the most adaptable, experiencing robust expansion within the market over recent years. Additionally, alternative storage technologies encompass compressed air and gravitational storage; however, their contribution to present-day power systems remains relatively modest [25].

Energy Storage Integration Trends

Several countries and regions have made significant strides in advancing their energy development goals. China led the way in 2022 with nearly 5 GW of grid-scale battery storage installations, while the United States closely followed with 4 GW, spurred on by the Inflation Reduction Act, which includes investment tax credits for standalone storage. India has also shown ambition, incorporating ambitious battery energy storage targets in its draft national electricity plan released in September 2022. Furthermore, the European Commission has taken action by issuing policy recommendations in March 2023 to promote the wider adoption of electricity storage across the European Union [25].

Pumped-storage hydropower retains its position as the most widely utilized storage technology, boasting a 2021 installed capacity of approximately 160 GW and a global capability of over 8,500 GWh in 2020, constituting more than 90% of the world's total electricity storage. The United States houses the largest capacity globally, primarily employed for daily balancing operations. Nevertheless, grid-scale batteries are rapidly gaining ground. Although currently smaller in capacity compared to pumped-storage hydropower, grid-scale batteries are projected to spearhead global storage growth, with close to 28 GW of installed capacity by the end of 2022, and more than 75% growth from 2021, with an addition of around 11 GW, as shown in Fig. 3.4. The United States and China led this expansion, each contributing gigawatt-scale installations. In terms of technology, lithium-ion batteries continued to dominate the grid-scale battery landscape in 2022, constituting the majority of newly added capacity [25].

While PHES is a pivotal technology for energy storage and grid stability, it potentially poses several environmental challenges [26, 27]. One significant side effect is the alteration of aquatic ecosystems due to the construction of large reservoirs, which can lead to habitat loss and fragmentation for various species. Additionally, the fluctuation of water levels in these reservoirs can disrupt the natural flow regimes of rivers, impacting sediment transport and water quality. The potential for increased evaporation rates from the surface of reservoirs can also exacerbate local water scarcity issues. Furthermore, creating and operating PHES facilities often necessitate substantial land use changes, potentially

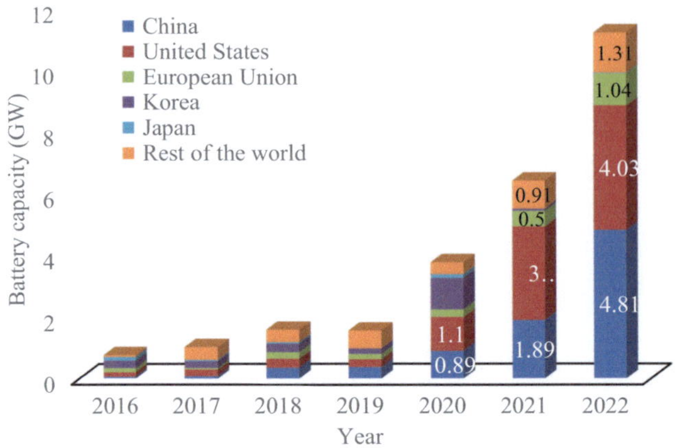

Fig. 3.4 Annual grid-scale battery storage additions (GW)

leading to deforestation, soil erosion, and negative impacts on local biodiversity [27]. These challenges should be taken into consideration when a new PHES is built. Also, detailed research needs to be conducted to quantify the side effects of PHESs.

The expansion of grid-scale battery storage is crucial, especially to align with the zero net emission scenario, which necessitates a substantial 35-fold increase in installed grid-scale battery capacity from 2022 to 2030, reaching nearly 970 GW, with a remarkable 170 GW added in 2030 alone, with the average of close to 120 GW per year from 2023 to 2030 [25].

Critical Minerals Demand for Energy Storage

While advancements in lithium-ion battery technology continue, achieving further cost reductions hinges on the pricing of critical minerals. Lithium iron phosphate batteries are the preferred choice for grid-scale storage due to cost and energy density considerations. In contrast, energy-dense lithium-ion chemistries like nickel cobalt aluminium (NCA) and nickel manganese cobalt (NMC) are gaining popularity in applications with limited space, such as home energy storage [21]. Beyond lithium-ion technology, flow batteries present a promising innovation for stationary storage. They exhibit minimal performance degradation over 25-30 years and offer flexibility in sizing according to specific energy storage needs, requiring limited initial investment. In 2022, China commissioned the world's largest vanadium redox flow battery with a capacity of 100 MW and a storage volume of 400 MWh [28].

While the past decade witnessed substantial price reductions in lithium-ion batteries, it is increasingly evident that further cost reductions rely not only on technological advancements but also on the trajectory of battery mineral prices. Lithium, a fundamental component of all lithium-ion batteries, remains a crucial factor in shaping the decarbonization of light passenger transport and energy storage. Additionally, the battery metals market has felt the impact of Russia's invasion of Ukraine, affecting the price of materials like nickel, cobalt and graphite, as Russia is the largest producer of battery-grade Class 1 nickel with a 20% market share [25]. Although 2022 marked the first increase in lithium-ion battery prices since 2010, with a 7% rise compared to 2021, some relief was observed only in the first quarter of 2023.

As seen in Fig. 3.5, the amount of minerals required for manufacturing energy storage and renewable technologies should increase 6 times by 2040 compared to 2020, which is a significant demand increase [29]. This level of demand is due to a higher level of mineral use in renewable technologies. For example, offshore wind turbines and solar PV need more than 15,000 and 6,800 kg/MW minerals, respectively, while a gas turbine needs only 1,100 kg/MW minerals, which is mainly copper [16]. Also, an electric car needs 7 times more minerals compared to a conventional car [16].

Crucial mining concerns demanding immediate attention encompass a range of pressing issues. These include the profound effects of mining activities on water resources, necessitating rigorous management and conservation measures. Additionally, the principle of obtaining free, prior, and informed consent from First Nations communities in each region must be central to mining endeavours, recognizing and respecting their rights and interests in these areas [30].

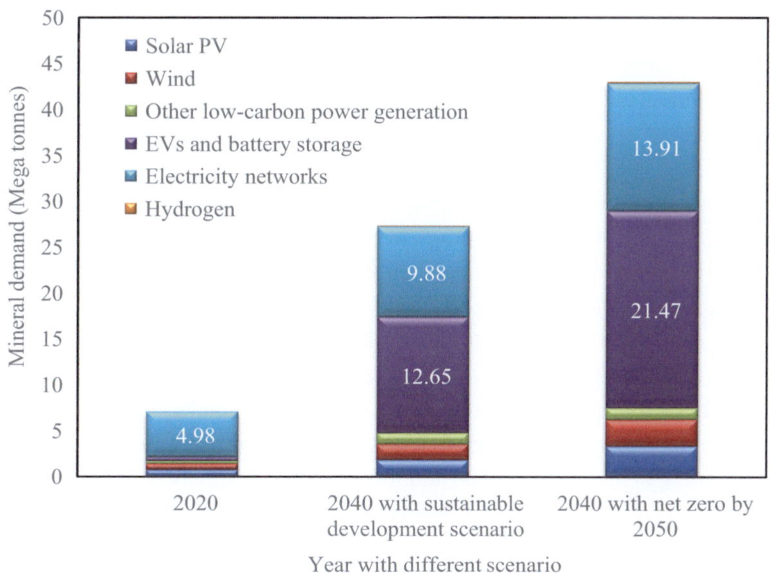

Fig. 3.5 Total mineral demand for clean energy technologies (Mt)

Furthermore, there is an essential call for integrated strategies that factor in climate change impacts, particularly extreme weather events that could disrupt the management of mine tailings. To ensure environmental sustainability, safeguarding biodiversity through comprehensive protection initiatives is imperative. Equally vital is the incorporation of thoughtful mine closure planning that seamlessly blends progressive ecosystem rehabilitation, leaving a lasting positive legacy. Finally, embracing circular business practices emerges as an essential approach, maximizing resource efficiency and minimizing waste within the mining industry [30].

Another challenge associated with energy storage is the shorter lifetime compared to other components of the energy systems. Chemical energy storage, which is usually based on lithium, has a lifetime of less than 20 years, but a traditional gas turbine is about 40-50 years. This issue makes energy storage more expensive. Flow batteries show a better lifetime of more than 25 years and offer a less expensive replacement, as only the liquid needs to be replaced after the end of the lifetime. However, still, the lifetime of chemical energy storage is not comparable with conventional energy producers. It is important to mention that in some implemented flow batteries, reasons for operational failures were identified [31–33] as follows. First, voltage variation from cell to cell can cause unequal charging. Second, membrane punctures allow water to leak between cells. Third, incorrect tubing arrangements lead to uneven electrolyte flow, resulting in poor charging and discharging behavior. Lastly, using unwashed electrodes after fixing tubing issues can introduce contaminants [31, 33]. All these reasons result in reducing the battery's efficiency, performance, and lifetime. For optimal flow battery operation, proper diagnostics and remedies for these issues are crucial. Some remedial methods include detecting leaks by circulating de-ionized water before electrochemical tests, feeding cells uniformly, identifying micro pinholes in membranes with half-cell leak tests, and adhering to the charging time duration and applied voltage [31, 32].

A critical area of research that warrants immediate attention is the seamless integration and effective control of energy storage systems within the electricity market. As the energy landscape continues to evolve, with a growing reliance on renewable sources like wind and solar, the role of energy storage becomes increasingly pivotal in stabilizing the grid and ensuring a reliable power supply. In this context, future research should focus on developing innovative grid management strategies that optimize the deployment of energy storage assets, ensuring they are strategically located to address grid congestion, enhance energy reliability, and support peak demand periods. Additionally, research efforts should delve into advanced control mechanisms, leveraging artificial intelligence and machine learning, to predict and respond to grid fluctuations in real-time, thereby maximizing the efficiency of energy storage systems and minimizing costs for both utilities and consumers. Ultimately, this research will be instrumental in facilitating a seamless transition towards a cleaner, more resilient, and economically efficient electricity market.

In addition, research is essential on efficient recycling of minerals from batteries and renewable technologies, which plays a pivotal role in the journey towards achieving net zero emissions, primarily due to the surging global demand for these critical resources. As the world shifts towards renewable energy sources and electric mobility, the need for minerals such as lithium, cobalt, and rare earth elements intensifies significantly. Proper recycling not only conserves these valuable resources but also reduces the environmental impact associated with mining and processing. Furthermore, recycling mitigates the risk of supply chain disruptions, ensuring a consistent and reliable source of materials crucial for the production of batteries and renewable technologies. By establishing robust recycling systems and implementing circular economy practices, we can significantly reduce the carbon footprint of these industries, contributing substantially to the overarching goal of achieving a sustainable and net-zero emission future.

References

1. T. L. Philip, Kai Li; Whitehead, Jake, presented at the Australasian Transport Research Forum (ATRF), Adelaide, SA, Australia, 28–30 September 2022, Driving and charging an EV in Australia: A real-world analysis. Available: https://espace.library.uq.edu.au/view/UQ:9c4d6d2.
2. C. Weiller, "Plug-in hybrid electric vehicle impacts on hourly electricity demand in the United States," *Energy Policy,* vol. 39, no. 6, pp. 3766–3778, 2011/06/01/ 2011.
3. O. Energy, "Origin Energy Electric Vehicles Smart Charging Trial Lessons Learnt," Origin Energy, Australia2022, Available: https://arena.gov.au/assets/2022/05/origin-energy-electric-vehicles-smart-charging-trial-lessons-learnt-2.pdf.
4. S. Mateen, M. Amir, A. Haque, and F. I. Bakhsh, "Ultra-fast charging of electric vehicles: A review of power electronics converter, grid stability and optimal battery consideration in multi-energy systems," *Sustainable Energy, Grids and Networks,* vol. 35, p. 101112, 2023/09/01/ 2023.
5. A. M. Omara, A. A. Shalabi, and M. R. Islam, "Voltage Stability Analysis of Grid-Connected Microgrid Incorporating Plug-in Electric Vehicles," in *2021 IEEE 6th International Conference on Computing, Communication and Automation (ICCCA)*, 2021, pp. 813–818.
6. M. R. Islam, H. Lu, M. J. Hossain, and L. Li, "Mitigating unbalance using distributed network reconfiguration techniques in distributed power generation grids with services for electric vehicles: A review," *Journal of Cleaner Production,* vol. 239, p. 117932, 2019/12/01/ 2019.
7. M. R. Islam, H. Lu, J. Hossain, M. R. Islam, and L. Li, "Multiobjective Optimization Technique for Mitigating Unbalance and Improving Voltage Considering Higher Penetration of Electric Vehicles and Distributed Generation," *IEEE Systems Journal,* vol. 14, no. 3, pp. 3676-3686, 2020.
8. K. M. Muttaqi, M. R. Islam, and D. Sutanto, "Future Power Distribution Grids: Integration of Renewable Energy, Energy Storage, Electric Vehicles, Superconductor, and Magnetic Bus," *IEEE Transactions on Applied Superconductivity,* vol. 29, no. 2, pp. 1-5, 2019.
9. V. T. Tran, M. R. Islam, D. Sutanto, and K. M. Muttaqi, "Reducing grid dependency of EV charging using renewable and storage systems," in Transportation, Cable Based and Wireless Charging Systems for Electric Vehicles: Technology and control, management and grid integration: Institution of Engineering and Technology, 2021, pp. 163–188. [Online]. Available: https://doi.org/10.1049/pbtr032e_ch6.
10. M. H. Shaker, H. Farzin, and E. Mashhour, "Joint planning of electric vehicle battery swapping stations and distribution grid with centralized charging," *Journal of Energy Storage,* vol. 58, p. 106455, 2023/02/01/ 2023.

11. J. Maleki Delarestaghi, A. Arefi, G. Ledwich, and A. Borghetti, "A distribution network planning model considering neighborhood energy trading," *Electric Power Systems Research,* vol. 191, p. 106894, 2021/02/01/ 2021.
12. S. LaMonaca and L. Ryan, "The state of play in electric vehicle charging services – A review of infrastructure provision, players, and policies," *Renewable and Sustainable Energy Reviews,* vol. 154, p. 111733, 2022/02/01/ 2022.
13. M. R. Islam, H. Lu, M. R. Islam, M. J. Hossain, and L. Li, "An IoT- Based Decision Support Tool for Improving the Performance of Smart Grids Connected With Distributed Energy Sources and Electric Vehicles," *IEEE Transactions on Industry Applications,* vol. 56, no. 4, pp. 4552-4562, 2020.
14. M. K. Hasan, A. K. M. A. Habib, Z. Shukur, F. Ibrahim, S. Islam, and M. A. Razzaque, "Review on cyber-physical and cyber-security system in smart grid: Standards, protocols, constraints, and recommendations," *Journal of Network and Computer Applications,* vol. 209, p. 103540, 2023/01/01/ 2023.
15. J. Buberger, A. Kersten, M. Kuder, R. Eckerle, T. Weyh, and T. Thiringer, "Total CO2-equivalent life-cycle emissions from commercially available passenger cars," *Renewable and Sustainable Energy Reviews,* vol. 159, p. 112158, 2022/05/01/ 2022.
16. IEA. (2021). *The Role of Critical Minerals in Clean Energy Transitions.* Available: https://www.iea.org/reports/the-role-of-critical-minerals-in-clean-energy-transitions.
17. AEMO. (2023). *Carbon Dioxide Equivalent Intensity Index.* Available: https://aemo.com.au/en/energy-systems/electricity/national-electricity-market-nem/market-operations/settlements-and-payments/settlements/carbon-dioxide-equivalent-intensity-index.
18. A. A. Ahmed, M. A. Nazzal, B. M. Darras, and I. M. Deiab, "Global warming potential, water footprint, and energy demand of shared autonomous electric vehicles incorporating circular economy practices," *Sustainable Production and Consumption,* vol. 36, pp. 449–462, 2023/03/01/ 2023.
19. Y. Li, M. K. Lim, Y. Tan, S. Y. Lee, and M.-L. Tseng, "Sharing economy to improve routing for urban logistics distribution using electric vehicles," *Resources, Conservation and Recycling,* vol. 153, p. 104585, 2020/02/01/ 2020.
20. X. Hu, Z. Yang, J. Sun, and Y. Zhang, "Sharing economy of electric vehicle private charge posts," *Transportation Research Part B: Methodological,* vol. 152, pp. 258–275, 2021/10/01/ 2021.
21. K. M. Tan, T. S. Babu, V. K. Ramachandaramurthy, P. Kasinathan, S. G. Solanki, and S. K. Raveendran, "Empowering smart grid: A comprehensive review of energy storage technology and application with renewable energy integration," *Journal of Energy Storage,* vol. 39, p. 102591, 2021/07/01/ 2021.
22. M. A. Hannan *et al.*, "Battery energy-storage system: A review of technologies, optimization objectives, constraints, approaches, and outstanding issues," *Journal of Energy Storage,* vol. 42, p. 103023, 2021/10/01/ 2021.
23. Y. Wu, T. Zhang, R. Gao, and C. Wu, "Portfolio planning of renewable energy with energy storage technologies for different applications from electricity grid," *Applied Energy,* vol. 287, p. 116562, 2021/04/01/ 2021.
24. L. A. Wong, V. K. Ramachandaramurthy, P. Taylor, J. B. Ekanayake, S. L. Walker, and S. Padmanaban, "Review on the optimal placement, sizing and control of an energy storage system in the distribution network," *Journal of Energy Storage,* vol. 21, pp. 489–504, 2019/02/01/ 2019.
25. IEA, "World Energy Outlook 2019," IEA, Paris 2019, Available: https://www.iea.org/reports/world-energy-outlook-2019.
26. S. Aryal, S. Ghimire, S. Tiwari, Y. Baaniya, and V. P. Pandey, "Evolution and future prospects of hydropower sector in Nepal: A review," *Heliyon,* vol. 10, no. 10, p. e31139, 2024/05/30/ 2024.

27. E. Gemechu and A. Kumar, "A review of how life cycle assessment has been used to assess the environmental impacts of hydropower energy," *Renewable and Sustainable Energy Reviews,* vol. 167, p. 112684, 2022/10/01/ 2022.
28. A. Colthorpe. (2022). *First phase of 800MWh world biggest flow battery commissioned in China.* Available: https://www.energy-storage.news/first-phase-of-800mwh-world-biggest-flow-battery-commissioned-in-china/.
29. IEA. *Total mineral demand for clean energy technologies by scenario, 2010-2040.* Available: https://www.iea.org/data-and-statistics/charts/total-mineral-demand-for-clean-energy-technologies-by-scenario-2010-2040-2.
30. A. government. (2023). *CRITICAL MINERALS STRATEGY 2023–2030.* Available: https://www.industry.gov.au/publications/critical-minerals-strategy-2023-2030/our-focus-areas.
31. R. Gundlapalli and S. Jayanti, "Case studies of operational failures of vanadium redox flow battery stacks, diagnoses and remedial actions," *Journal of Energy Storage,* vol. 33, p. 102078, 2021/01/01/ 2021.
32. J. Xiong, M. Jing, A. Tang, X. Fan, J. Liu, and C. Yan, "Mechanical modelling and simulation analyses of stress distribution and material failure for vanadium redox flow battery," *Journal of Energy Storage,* vol. 15, pp. 133–144, 2018/02/01/ 2018.
33. F. Reichelt and K. Müller, "Assessment of the reliability of vanadium-redox flow batteries," *Engineering Reports,* vol. 2, no. 10, p. e12254, 2020.

Net-Zero Emission Energy System Planning and Circular Economy

4

'Net-zero emissions' refers to achieving a balance between the greenhouse gas emissions produced and those removed from the atmosphere. The transition towards a zero net emission energy system presents significant challenges in terms of planning and investment decisions, especially within the electricity market. One of the critical approaches towards zero net emissions is the change of economic concepts. As the world's population continues to grow and living standards improve, there is an increasing demand for essential resources such as food and energy. Simultaneously, the widespread adoption of manufacturing using the linear economy model of take-make-dispose, in which there is no or little recycling, has resulted in a surge in waste generation. This linear approach to the economy has not only caused environmental and public health risks due to resource depletion and waste production but has also triggered a call for change. Enter the circular economy concept, a closed-loop, waste-free model gaining traction in the global industry.

Net-Zero Emission Energy System Planning Concept

A report by the International Monetary Fund (IMF) for the Group of 20 states that to keep global temperatures in control, governments must swiftly implement comprehensive climate policies [1]. A growth-oriented strategy comprises four key components: a green investment plan, carbon pricing, support for green research and development (R&D), and measures to ensure a fair societal transition. Green investments would lead to phased-in increases in carbon prices while boosting demand and employment. Carbon pricing, either through taxes or emission trading systems, is essential for substantial emission reductions, complementing green investments and R&D. This study also suggests that

Md. R. Islam et al., *Universal Access to Sustainable, Affordable, and Reliable Electricity*,
Synthesis Lectures on Sustainable Development,
https://doi.org/10.1007/978-3-032-19517-3_4

aligning infrastructure with net-zero emissions necessitates public investments of 0.5 to 4.5% of GDP over a decade, mainly shifting focus from fossil fuels to renewables, electrification, and energy efficiency. Public investment plays a crucial role alongside private sector contributions. Delaying carbon pricing could jeopardize net-zero emission targets, causing irreversible climate and economic damage. In challenging-to-decarbonize sectors, sector-specific policies should complement carbon pricing.

The transition to net-zero emissions must be equitable, addressing employment shifts, especially in coal regions, through retraining, geographic mobility support, and fostering new industries. Carbon pricing revenues can compensate households for increased energy costs, while developing economies benefit from basic infrastructure improvements. Global cooperation, including technology transfer and climate finance for developing nations, is vital. A coordinated green investment push involving all countries, differentiated by development level, can boost demand, limit carbon leakage, and enforce a global carbon price floor. To achieve this, financial and technological support for developing economies is also essential, considering their energy demands and limited fiscal resources.

Investment and Energy Demand Implications in Net-Zero Scenarios

It is expected that the world invests between 2.2 and 3.2% of the GDP for the net zero emission target, as shown in Fig. 4.1. The equivalent amount of investment is about US$244 trillion, which would be a significant amount of investment by 2050. This figure is very critical to revisit when it is reported [1] that total employment number due to this plan will be reduced for almost three decades, with the same number of employment in 2050 as it was in 2028.

The study by the net-zero Australia (NZAu) [2] shows that for achieving net-zero energy emissions by 2050, Australia must embark on a comprehensive energy transformation. This transformation involves several key initiatives:

- Renewable Energy Growth: Australia needs to significantly expand its reliance on renewable energy sources, aiming for 400–500 GW of capacity by mid-century for domestic use and potentially several thousand GW for energy exports. This represents a substantial increase from the current 25 GW capacity.
- Energy Storage: A fleet of batteries, pumped hydro energy storage, and gas-fired firming systems larger in capacity than the existing domestic energy system is required to stabilize the grid and ensure a reliable energy supply.
- Electrification and Efficiency: To reduce emissions across all sectors, Australia must promote electrification and enhance energy efficiency. This includes greater adoption of electric vehicles and replacing gas-fired heating with electric alternatives in homes and businesses.

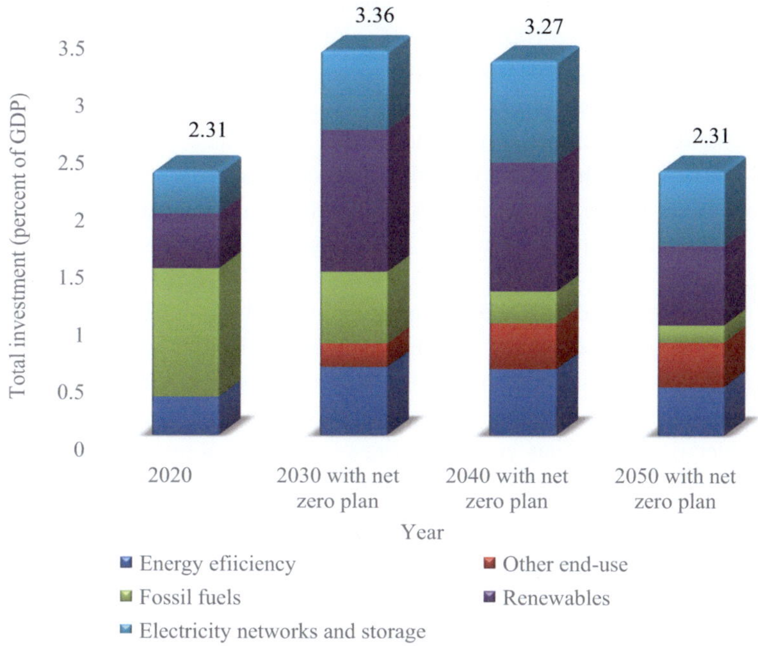

Fig. 4.1 World total investment, percent of GDP; in annual average per decade to achieve zero net emission

- Carbon Capture and Storage (CCS): The development of a substantial CCS industry capable of sequestering around 100 million tons of CO_2 annually is crucial. This represents significant growth from the current nascent CCS industry.
- Infrastructure Expansion: Expansion of energy networks is necessary, not only through electrification but also by investing in new infrastructure for delivering various commodities.
- Capital Attraction: Australia needs to attract a substantial capital investment of $7–9 trillion by 2060 from both international and domestic sources to fund projects that support the energy transformation.

Additionally, transforming Australia's exports is integral to global decarbonization efforts. This shift entails transitioning to clean energy and clean mineral exports, potentially involving thousands of GW in renewable and other investments, especially in northern Australia. Clean production of iron and aluminum onshore could offer cost advantages over offshore production, aligning with global decarbonization goals and international demand.

Investment in the workforce is also essential, with the need to expand from approximately 100,000 jobs today to 700,000–850,000 jobs by 2060. Many of these employment opportunities will be in regional and remote areas.

Australia must also focus on reducing the adverse impacts of the net-zero transition and ensuring benefits are shared. This includes moving the land sector towards net-zero or potentially net-negative emissions, requiring innovative approaches as offsets from land use may not be available for other sectors. Careful management of land use changes is critical, with particular consideration for Indigenous land, natural ecosystems, and agriculture.

In this study [2], six scenarios are considered, such as the reference scenario (REF), which is a business-as-usual case, and the rapid electrification scenario (E +). The E + Scenario envisions a comprehensive shift towards electrification in the transport and building sectors, aiming for near-complete electrification by the year 2050. This transition entails electrifying various energy services in residential and commercial buildings, encompassing functions such as air conditioning, space heating, ventilation, water heating, lighting, refrigeration, freezing, clothes washing and drying, dishwashing, and cooking. The adoption of these electrified services follows a trajectory resembling an S-Curve, with the goal of achieving full saturation of electrified building services across building stocks by the year 2050. In the transport sector, the scenario distinguishes between light-duty vehicles (LDVs), heavy-duty vehicles (HDVs), and buses. By the year 2040, it is projected that 88% of LDV sales will comprise battery electric vehicles (BEVs), while 11% will consist of hydrogen fuel cell vehicles (HFCVs). In the case of HDVs, it is anticipated that 65% of sales will be BEVs, with 34% being HFCVs. Similar to building electrification, the adoption of BEVs and HFCVs in the transport sector is expected to follow an S-Curve trajectory, with the aim of achieving full saturation of these technologies in the transport sector by around the year 2050.

As shown in Fig. 4.2, about 30% of final domestic energy demand is from non-green energy resources such as Gasoline, Diesel, Oil & its products, Brown coal, and Black coal in E + scenario. However, the scenario allows for geological CO_2 sequestration for achieving net-zero emissions. In the REF scenario, about 75% of the final domestic demand is associated with non-green energy resources. Another assumption in this study is that the demand in E + scenario is reduced by 40% by 2060, compared to the total demand in the REF scenario due to electrification, energy efficiency and productivity measures. However, key technologies such as peer-to-peer transactions for maximizing the benefits of PV and batteries for local neighbors are not considered in this study.

In terms of technology, as seen in Fig. 4.3, there is still a plan for building more OCGT and CCGT and the use of coal. Large-scale solar and wind will be the main technologies forecasted in this study. However, some emerging technologies, such as nuclear power based on small modular reactors are or direct transfer of energy from space, are not considered in this study.

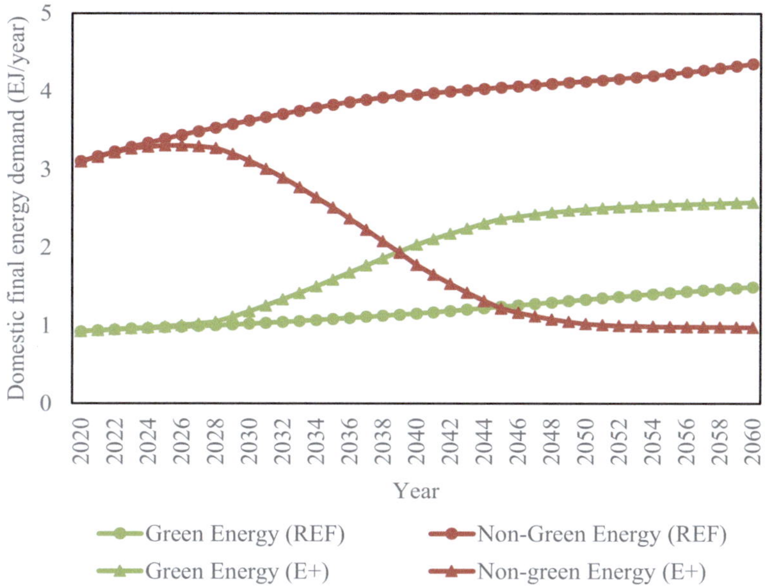

Fig. 4.2 Domestic final energy demand (EJ/year)

The Impact of Zero-Net Targets on Material Demand

Another challenge with respect to the zero-net emissions energy is the required material to make technologies for this purpose. To achieve zero net energy by 2050, a substantial quantity of materials will be required. The study's findings underscore a noteworthy disparity between the anticipated material needs in 2050 and the feasible net zero production in 2050, taking into account resource limitations. Specifically, this disparity could amount to 300–930 million tons for steel, 155–360 million tons for paper, and up to 30–55 million tons for aluminium [3].

The findings of a study [4] indicate a significant increase in material requirements. Notably, there is a surge in demand for Indium (In), Gallium (Ga), and Arsenic (As), driven by the introduction of GaAs PV cells in the market post-2030. For most other materials, peak demand is projected to occur in the mid-2030s. Furthermore, it is anticipated that silver's supply will surpass primary demand after 2040. Also, there will be an increased demand for Tellurium (Te), Cadmium (Cd), and Germanium (Ge) until 2040. The rapid expansion of wind capacity and the widespread adoption of electric vehicles (EVs) contribute to a sustained rise in demand for rare earth elements (REEs). In the 2040–2050 period, REE demand for NdFeB magnets is predicted to increase significantly, primarily due to electric motors (EMs). EMs will account for approximately 70% of the total REE demand from 2021 to 2050, compared to only 17% in the preceding decade

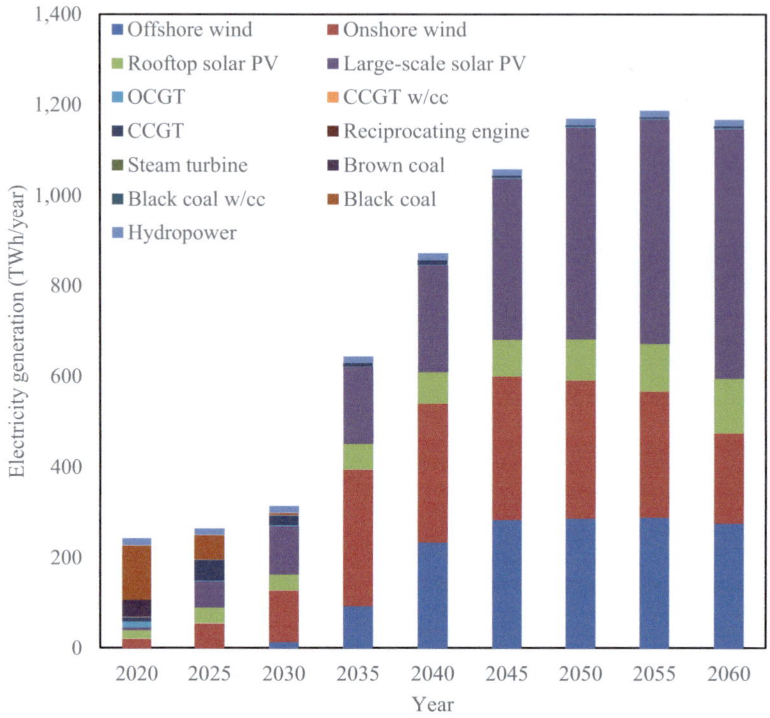

Fig. 4.3 Projected domestic electricity generation, by technology (TWh/year)

(2011–2020). The demand for various cathode materials is expected to grow continuously due to the widespread adoption of new EVs, with battery chemistry influencing material requirements. Advances in battery technology, like NMC811 or advanced solid-state batteries (ASSBs), will reduce the Material Intensity of Cobalt (Co) and Manganese (Mn) but increase the demand for Lithium (Li) and Nickel (Ni).

However, refining, extracting, and processing the above-mentioned material needs an enormous amount of mining and energy resources, which affects environmental sustainability in another way. A comprehensive study needs to be conducted to consider the impact of such significant demand on the material, including REEs.

Another aspect to be considered very carefully is the effective design of the distribution and transmission grid to deliver electricity to load points. In Australia, it is calculated that more than A$335 billion is required to invest in the grid to support the zero-net emission goal in the scenario E + [2]. However, in this study, key technologies and strategies such as superconducting cable deployment in congested areas, and power electronic transformers for enhanced flexibility are not considered. In addition, the grid needs a great amount of supporting equipment, such as a synchronous condenser, to keep the grid stability within the required levels. However, these stability analyses have not been conducted to evaluate realistic costs associated with the zero net emission plan.

In conclusion, we should explore a novel model for zero-net emissions by minimum investment in energy systems, minimum impact on environmental sustainability, considering extra mining for elements, and maximizing the empowerment of the employment rate, aiming to achieve zero net emissions. In such a model, we should be considering key technologies and strategies such as superconducting cable deployment in congested areas, power electronic transformers for enhanced flexibility, peer-to-peer transactions, demand management, and energy storage. By incorporating these elements into the planning process, the proposed model aims to optimize the cost-effectiveness and sustainability of the energy system. Without proper planning, the cost of achieving zero net emissions will be much more than the amount that is estimated currently.

Circular Economy Concept

The innovative approach of the circular economy has the potential to safeguard the environment and boost economic well-being by reducing resource and energy consumption. To make this shift successful, it's crucial to focus on key elements such as green technologies, decarbonization strategies, bio refinery processes, material flow analysis, life cycle assessment, ecological footprints, substance flow analysis, circularity index, eco-design, bioresource management, new business models, and policy development [5–9]. Collaboration and partnerships among government entities, stakeholders, policymakers, and competent authorities are also vital to establishing a circular ecosystem, fostering socio-economic sustainability, ecological preservation, and industrial progress for a sustainable future.

To establish an effective circular economy, key strategies encompass a range of practices aimed at prolonging product life, including reuse, repair, repurpose, refurbish, recondition, upgrade, retrofit, redesign, and remanufacture. These strategies are complemented by recycling efforts focused on closing resource cycles [8]. Central to the success of the circular economy is the pivotal role played by de-manufacturing processes. These processes necessitate the implementation of a comprehensive end-of-life product recovery system, involving thorough inspections and evaluations of product integrity, ultimately guiding them into processing cycles for reintroduction into production processes. The efficiency of this reverse chain is paramount for the model's effectiveness, as the widespread use of recycled materials hinges on proper management [9].

Transitioning to circular manufacturing requires substantial initial investments in areas like product design, new machinery, production lines, and supply chain restructuring. The primary aim of circular materials is to minimize or eliminate potential harm to human health and the environment, and the selection of suitable materials is a complex task, given the diverse materials often used in products and their global origins. Reorganizing supply networks for circular systems is also challenging, especially in today's globalized markets, where material extraction, manufacturing, and consumption occur across different countries and continents, making fully closed-loop systems difficult to achieve [10].

Another obstacle in adopting a circular economy is the higher cost of producing durable products compared to single-use alternatives. To break away from the linear economic model, where environmental and health costs are separated from production and consumption, these costs need to be integrated into consumer prices. Additionally, governments' financial incentives and access to bank financing for the implementation of a circular economy should be facilitated [9].

Circular Economy for Battery

Recycling or repurposing batteries can reduce the GHG emissions of batteries. In the research conducted in [11], real data are used to calculate the emission savings associated with battery recycling or repurposing. Battery recycling means that the battery is processed thoroughly for extracting minerals from the battery, while repurposing means that the battery is refurbished to be used in another application in order to increase the useful lifetime of the battery. In this study [11], the transportation of batteries to the recycling plant, fuel consumption efficiency, and loading capacity of those transportation in different transportation routes are also considered. Also, the amount, type, and emissions associated with the energy consumed in the battery recovery plant are also considered.

To quantify the GHG savings associated with recycling lithium-ion batteries, a comparison was made between the life-cycle GHG emissions resulting from the recycling process and the GHG emissions generated by the manufacturing of the same value of batteries and equivalent volumes of raw materials [11]. In essence, this analysis sought to assess the environmental benefits of recycling lithium-ion batteries by evaluating how they compare to the GHG emissions produced when creating new batteries and the necessary raw materials from scratch.

As seen in Fig. 4.4, GHG savings from repurposing and recycling lithium-ion batteries are 651.98 and 828.38 kg CO_2-eq/tonnes, respectively. This amount is about 5% of the total emission produced during the production of Li-ion batteries, as shown in this Figure. In this Figure, the greenhouse gas emissions of lithium-ion battery per kWh production is about 65 kg CO_2-eq in 2020 and considering the fact that the energy density is around 250 Wh/kg.

Therefore, recycling and repurposing of batteries are crucial when planning for a transition to zero-net emission society. Research needs to be conducted to provide a model and value chain for the circular economy in battery and energy storage lifetime. The recovery of materials from end-of-life batteries is projected to increase after 2030. A low

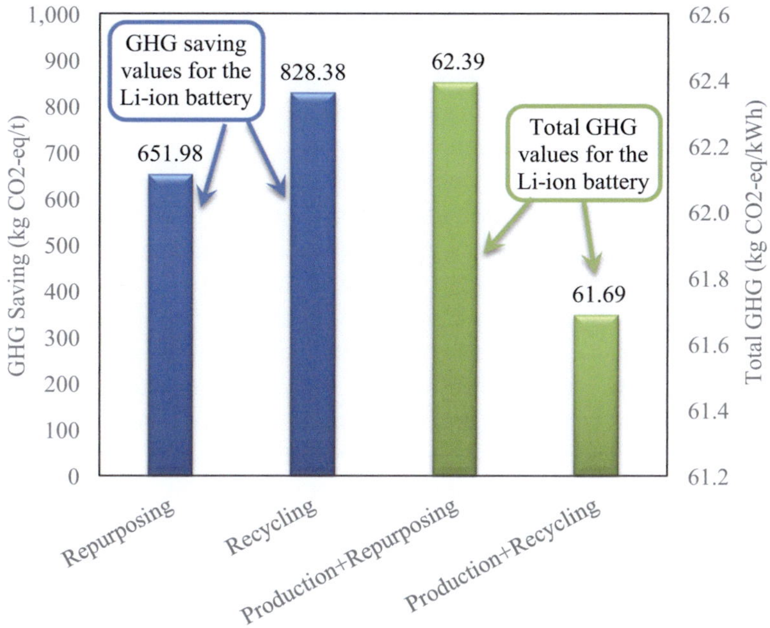

Fig. 4.4 GHG equivalent of Lithium-ion battery repurposing and recycling, and total GHG of Lithium-ion battery, including repurposing and recycling

recycling rate of 10% in 2050 is expected to meet less than 7% of the total demand, but an 80% recycling rate could fulfil around 50% of the primary demand.

Circular Economy in Power System

In the context of power systems, the circular economy principles aim to optimize the lifecycle of equipment by promoting sharing, leasing, reusing, repairing, refurbishing, and recycling existing materials and products. This paradigm shift towards a circular economy in power systems presents opportunities to minimize waste, reduce resource consumption, and enhance sustainability. The application of circular economy principles and new business models in power systems should be evaluated, examining the challenges, benefits, and potential strategies for implementing circularity in equipment, knowledge, and service management. This can include the study on extending the lifetime of existing network components such as transformers and switchgear [12].

Some components used in transformers and cables, such as copper and aluminium are not degraded and can be recycled infinitely. This is a critical rule to consider when eco-designing the products and planning for long-term in power systems. Copper windings are recycled into three distinct purity levels: Purity 1A, 1B, and 2. The highest level,

Purity 1A (> 99.9%), matches the quality of copper cathodes and can be directly melted and reused in electrical applications like transformer windings, creating a highly efficient circular route in terms of cost, energy use, and environmental impact. Purity level 1B (> 99.7%) shares quality with copper anodes and requires an electrolysis process to achieve the necessary purity for electrical copper. Purity level 2 (> 95.0%) is akin in quality to copper concentrate sold by mining companies, necessitating a thermo-chemical process to reach level 1B and then an electrolysis process to reach level 1A. However, in practice, it is often used without further purification in applications that demand lower purity, such as copper alloys (e.g., brass) or as additives in steel manufacturing. In theory, aluminium wires has the potential for recycling at purity levels of up to 99.7% (Alu 7), meeting the required standard for reuse in the manufacture of new aluminium electrical wires [13].

Transformer oil is theoretically possible to be recovered completely, however, the mineral transformer oil through filtration can be extracted approximately 70% as first-grade oil for reuse in transformers, with the remaining 30% finding use as second-grade oil in various industrial applications, primarily serving as an industrial lubricant. In the case of natural esters, the potential for recovery exists similarly to mineral oil, but a first-grade recycling process for natural esters has yet to be established due to the absence of a comprehensive market, suitable equipment, and technical standards. Presently, an estimated 20–30% of natural ester insulation undergoes second-grade recycling as vegetable oil, while the remainder is incinerated with heat recovery [13].

Nearly 100% of magnetic steel extracted from the transformer core and structural steel components is successfully recovered. While magnetic steel laminates can theoretically be cleaned and reused in new transformers, sometimes, they need to be reprocessed due to new standards and the eco-design of transformers. Where there are no such restrictions, the process involves stamping out smaller laminates from the original piece, eliminating concerns about potential damage at the laminate edges. The proportion of magnetic steel taking this route depends on prevailing prices in the Asian market [13, 14].

As shown in Fig. 4.5, more than 75% of a typical transformer can be recycled to the first-grade material for reusing in transformers or other applications in power system engineering. Therefore, in the path to energy transition, it is required for all manufacturers to declare how much they are using recycled material and what the incremental rate of using such elements is.

Considering that the integration of high levels of PV and demand response impacts the lifetime of equipment such as transformers [15], it is critical to establish a comprehensive preventive maintenance program on transformers to extend their useful lifetime as much as practical [16, 17].

Switchgears are another widely used component in power systems. Increasing the lifetime of switchgear, reducing the maintenance cost and planning for recycling from the beginning need to be integrated in all processes of the engineering, design, and manufacturing. To enhance the lifespan of switchgear, several approaches are adopted. Firstly, the switchgear's primary system tank is hermetically sealed to prevent moisture and dust

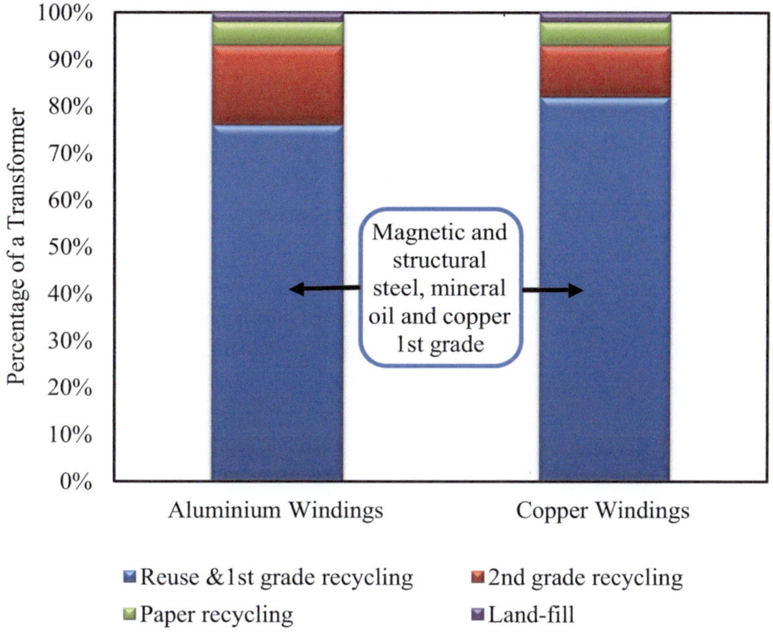

Fig. 4.5 Circularity opportunity in a 400 kVA, mineral oil filled transformer

infiltration, ensuring robustness without overpressure, and utilizing long-lasting circuit breakers instead of alternative fuses. Moreover, employing a self-powered protection relay in conjunction with a Circuit Breaker design provides significant improvement in the reliability of switchgear. This combination yields benefits such as minimal energy losses, elimination of replacement costs, and reduced inspection and maintenance requirements [18].

A new model for the life-cycle assessment of emissions, including circular economy for power system components, needs to be developed in order to evaluate the true value of circularity, and also to ensure eco-design of such components. In other words, the components should be designed *for more than one life*! To address the circularity of equipment, it is suggested to provide a circularity grade or label for each component. For example, the following criteria can be considered to provide circularity or eco-friendly grades to equipment [19]:

- Life-cycle carbon footprint: Measured from the perspective of the emission through all processes and integrated with other factors to provide a comprehensive life-cycle carbon footprint assessment.
- Substances of very high concern: Ensures the absence of carcinogenic, mutagenic, reproductive toxic, or environmentally hazardous substances in equipment.

- Recyclability/circularity: Emphasizes the potential recyclability of materials used in equipment or the full recyclability of the products themselves.
- Recycling input rate: considers the presence of recycled materials in equipment.
- Environmental benefits: recognizes low-carbon enabling products, regulation compliant equipment, and those used for green energy sources as environmentally advantageous.
- Equipment transmission efficiency: Evaluates the sustainability of equipment performance, with higher efficiency indicating greater sustainability.

A study also shows that the GHG of a wind turbine with and without circular economy would be 499 and 623 tonnes CO_2 equivalent, revealing a 20% reduction in GHG due to circularity [20]. In terms of electric motors, a significant gap is uncovered in the methodology for selecting optimal end-of-life approaches for industrial electric motors, including reusing, refurbishing, and recycling, due to limited information about product conditions and availability. The survey findings within the UK indicated that nearly half of these companies do not currently employ repair strategies for electrical machine components. However, there was a notable desire among respondents to transition their businesses toward more sustainable practices. The growing volume of End-of-Life Permanent Magnets is likely to stimulate a recycling industry. While current recycling rates are projected to provide only a small fraction of the primary demand (<1%), a high recycling rate scenario (e.g., 100% recovery from wind turbines and 80% from motors) could satisfy up to 35% of the primary demand by 2050 [21]. The research should answer how many percentages of new materials for equipment manufacturing should come from recycling and what the trend of this percentage is from now to 2050.

Countries with limited recycling capabilities can engage in a circular economy by promoting reuse and repair, encouraging collaborative consumption, and leveraging international partnerships. By fostering initiatives like repair cafes, second-hand markets, and upcycling projects [22], these countries can extend product lifecycles and reduce waste. Embracing shared economy models, such as car and food sharing, tool libraries, and rental services, can lower the need for individual ownership and decrease the overall material footprint [23]. Additionally, collaborating with international organizations and other countries can provide an opportunity of investing in other countries for recycling and waste management plants and then exporting waste to those countries for processing. Detailed research needs to be conducted to explore how countries with limited recycling capabilities can manage and process waste and promote a circular economy.

References

1. (2022). *Australia joins international Net Zero Government Initiative.* Available: https://www.dcceew.gov.au/about/news/australia-joins-international-net-zero-government-initiative.

2. Z. N. Australia, "Final modelling results," 2023, Available: https://www.netzeroaustralia.net.au/wp-content/uploads/2023/04/Net-Zero-Australia-final-results-full-results-pack-19-April-23.pdf.
3. L. Gast and J. M. Allwood, "What bulk material production is possible on a transition to net zero emissions by 2050 with limited zero emissions resources?," *Journal of Cleaner Production*, p. 138346, 2023/08/18/ 2023.
4. Y. Liang, R. Kleijn, and E. van der Voet, "Increase in demand for critical materials under IEA Net-Zero emission by 2050 scenario," *Applied Energy*, vol. 346, p. 121400, 2023/09/15/ 2023.
5. R. A. Bellezoni, A. P. Adeogun, M. X. Paes, and J. A. P. de Oliveira, "Tackling climate change through circular economy in cities," *Journal of Cleaner Production*, vol. 381, p. 135126, 2022/12/25/ 2022.
6. E. Sonnier and A. Grit, "A narrative for circular economy in Cities: Conditions for a Mission-Oriented innovative system," *City and Environment Interactions*, vol. 16, p. 100084, 2022/12/01/ 2022.
7. P. K. Mukherjee *et al.*, "Socio-economic sustainability with circular economy — An alternative approach," *Science of The Total Environment*, vol. 904, p. 166630, 2023/12/15/ 2023.
8. M. Demartini, M. Ferrari, K. Govindan, and F. Tonelli, "The transition to electric vehicles and a net zero economy: A model based on circular economy, stakeholder theory, and system thinking approach," *Journal of Cleaner Production*, vol. 410, p. 137031, 2023/07/15/ 2023.
9. P. S. Varbanov *et al.*, "Efficiency measures for energy supply and use aiming for a clean circular economy," *Energy*, p. 129035, 2023/09/17/ 2023.
10. M. Milousi and M. Souliotis, "A circular economy approach to residential solar thermal systems," *Renewable Energy*, vol. 207, pp. 242–252, 2023/05/01/ 2023.
11. X. Zhang, Y. He, Y. Wang, W. Yan, and N. Subramanian, "Assessing the GHG emissions and savings during the recycling of NMC lithium-ion batteries used in electric vehicles in China," *Processes*, vol. 10, no. 2. https://doi.org/10.3390/pr10020342.
12. IBM, "The future of energy business in a circular economy," IBM, New Orchard Road, Armonk, NY 105042022, Available: https://www.ibm.com/downloads/cas/5WKLAP2D.
13. T. J. Bruno De Wachter. (2022) The circularity of medium-power electrical transformers: Evaluating design options based on an in-depth market survey. *TRANSFORMERS MAGAZINE*. 84–89. Available: https://hrcak.srce.hr/file/392582.
14. A. Fontana *et al.*, "Circular economy strategies for equipment lifetime extension: A systematic review," *Sustainability*, vol. 13, no. 3. https://doi.org/10.3390/su13031117.
15. B. Behi, A. Arefi, H. Pezeshki, and F. Shahnia, "Distribution transformer lifetime analysis in the presence of demand response and rooftop PV integration," *Renew. Energy Environ. Sustain.*, vol. 2, p. 27, 2017.
16. B. Nemeth, T. Benyo, A. Jager, G. Csepes, and G. Woynarovich, "Complex diagnostic methods for lifetime extension of power transformers," in *Conference Record of the 2008 IEEE International Symposium on Electrical Insulation*, 2008, pp. 132–135.
17. M. Ciontu and M. A. Boruz, "Preventive maintenance optimization for power transformers in use," in *2012 International Conference on Applied and Theoretical Electricity (ICATE)*, 2012, pp. 1–5.
18. EATON, "The circular economy: A case study of design considerations for medium voltage switchgear," Switzerland2019, Available: https://www.eaton.com/content/dam/eaton/products/medium-voltage-power-distribution-control-systems/sf6-free-switchgear/eaton-wp-circular-rmus-white-paper-en-gb.pdf.
19. P. Group. (2023). *ECO CABLE, green inside*. Available: https://na.prysmiangroup.com/sustainability/eco-cable.

20. B. Diez-Cañamero and J. M. F. Mendoza, "Circular economy performance and carbon footprint of wind turbine blade waste management alternatives," *Waste Management,* vol. 164, pp. 94–105, 2023/06/01/ 2023.
21. D. Tiwari, J. Miscandlon, A. Tiwari, and G. W. Jewell, "A Review of Circular Economy Research for Electric Motors and the Role of Industry 4.0 Technologies," *Sustainability*, vol. 13, no. 17. https://doi.org/10.3390/su13179668.
22. W. Spekkink, M. Rödl, and M. Charter, "Repair Cafés and Precious Plastic as translocal networks for the circular economy," *Journal of Cleaner Production,* vol. 380, p. 135125, 2022/12/20/ 2022.
23. M. Ranjbari, Z. Shams Esfandabadi, P.-O. Siebers, P. Pisano, and F. Quatraro, "Digitally enabled food sharing platforms towards effective waste management in a circular economy: A system dynamics simulation model," *Technovation,* vol. 130, p. 102939, 2024/02/01/ 2024.

Nuclear Power

5

Nuclear energy has long occupied a complex position in global debates on electricity generation, offering the promise of large-scale, low-carbon power while raising concerns regarding safety, cost, and long-term waste management. In 2020, nuclear power comprised 10% of the world's electricity generation, a decline from the late 1990s when it represented 18%. Despite this decrease, nuclear energy remains the second-largest source of low emissions electricity, following hydroelectricity, and is the primary source in advanced economies. This surpasses the combined generation from wind and solar PV worldwide in 2020, demonstrating its continued significance [1]. This chapter examines the evolving role of nuclear technologies in the context of net-zero emission targets, focusing on both conventional reactors and the emergence of small modular reactors (SMRs) as potentially transformative solutions. It discusses how nuclear power contributes to emission reduction efforts by displacing fossil fuel generation, while also addressing the economic realities of high capital costs and the social and political challenges surrounding public acceptance. Particular attention is given to the technological innovations and operational flexibility offered by SMRs, which are increasingly seen as a means of enabling safer, more adaptable nuclear deployment. By exploring these opportunities and constraints, the chapter situates nuclear power within the broader portfolio of sustainable energy strategies, emphasizing its potential role as a complementary source alongside renewable resources.

Md. R. Islam et al., *Universal Access to Sustainable, Affordable, and Reliable Electricity*,
Synthesis Lectures on Sustainable Development,
https://doi.org/10.1007/978-3-032-19517-3_5

Nuclear Power Installations and Emissions

As of the end of 2021, there were 439 operational nuclear power reactors in 32 countries, boasting a collective capacity of 413 GW, with approximately 270 GW located in advanced economies. As shown in Fig. 5.1, nuclear power has significantly contributed to reducing global CO_2 emissions since the 1970s, preventing approximately 70 Gt of CO_2 emissions between 1971 and 2020. These emissions reductions were most prominent in advanced economies, with over 85% of avoided emissions in the European Union (20 Gt) and the United States (24 Gt). Without nuclear power, emissions from electricity generation would have been substantially higher in Japan and substantially higher in Korea and Canada, making it a crucial component in the battle against climate change [2].

When considering GHG emissions on a lifecycle basis, it's evident that coal-fired power plants top the list with the highest emission intensities. On the contrary, biomass, nuclear, hydroelectric, wind, and solar photovoltaic energy sources exhibit significantly lower GHG emissions compared to fossil fuels. It's essential to acknowledge that constructing nuclear power plants tends to result in higher greenhouse gas emissions compared to the construction of other electricity generation systems, a factor to consider when evaluating their environmental impact. This is due to the fact that nuclear power plants

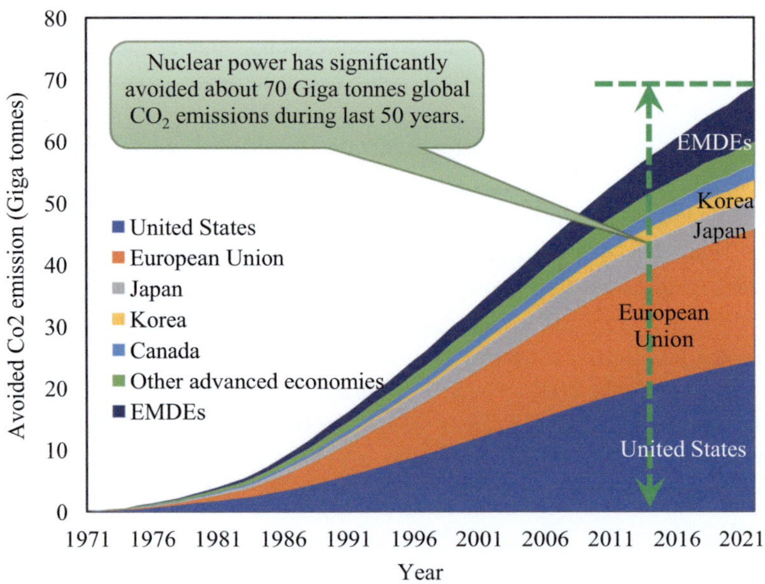

Fig. 5.1 Avoided CO_2 emission by nuclear power from 1970 to 2022, EMDEs = emerging market and developing economies including China, Advanced economies consist of Australia, Canada, Chile, the 28 members of the European Union, Iceland, Israel, Japan, Korea, Mexico, New Zealand, Norway, Switzerland, Turkey, and the United States

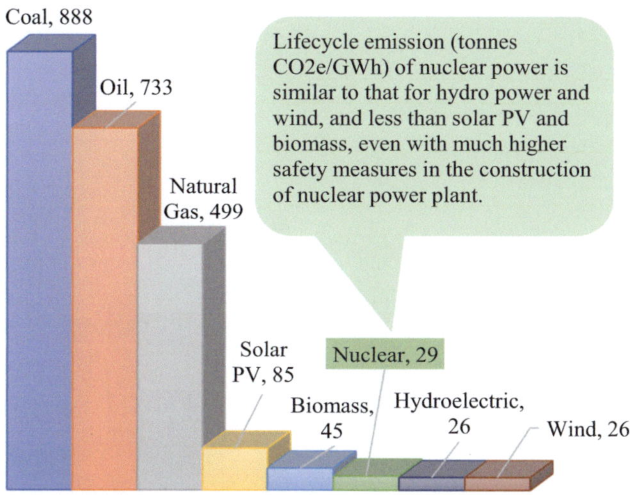

Fig. 5.2 Lifecycle emission of power generation by different technologies (tonnes CO_2e/GWh)

ensure a high level of safety through the defense-in-depth approach, incorporating multiple physical barriers in their design [3]. Figure 5.2 illustrate the lifecycle emission of power generation by different technologies in tonnes CO_2e/GWh.

Nuclear Power Costs

Towards a zero-net emissions strategy by 2050, nuclear power can play a vital role in reducing emissions and providing lower cost electricity. In 2022, nuclear power capacity experienced significant growth, with 7.9 GW of new capacity added, representing a 40% increase compared to the previous year. China continued its remarkable trend of leading in nuclear power expansion for the tenth consecutive year, completing two reactors. Additionally, projects were successfully finished in several countries, including Finland, Korea, Pakistan, and the United Arab Emirates. Also, 2022 also marked the commencement of construction on several reactors, with five underway in China, two in Egypt, and one in Turkey, which shows the global commitment to expanding nuclear energy [4]. Nuclear power continues to stand out as a reliable, low-carbon energy source with expected competitive pricing in 2025, surpassing fossil fuel-based generation. While large hydro reservoirs offer a similar contribution, they are contingent on natural resources. In contrast, nuclear plants are expected to be more cost-effective than many types of technologies in terms of the levelized cost of electricity (LCOE). Although gas-based combined-cycle gas turbines (CCGTs) are competitive in certain regions, their LCOEs hinge on gas and carbon emission prices. Long-term operation (LTO) through lifetime

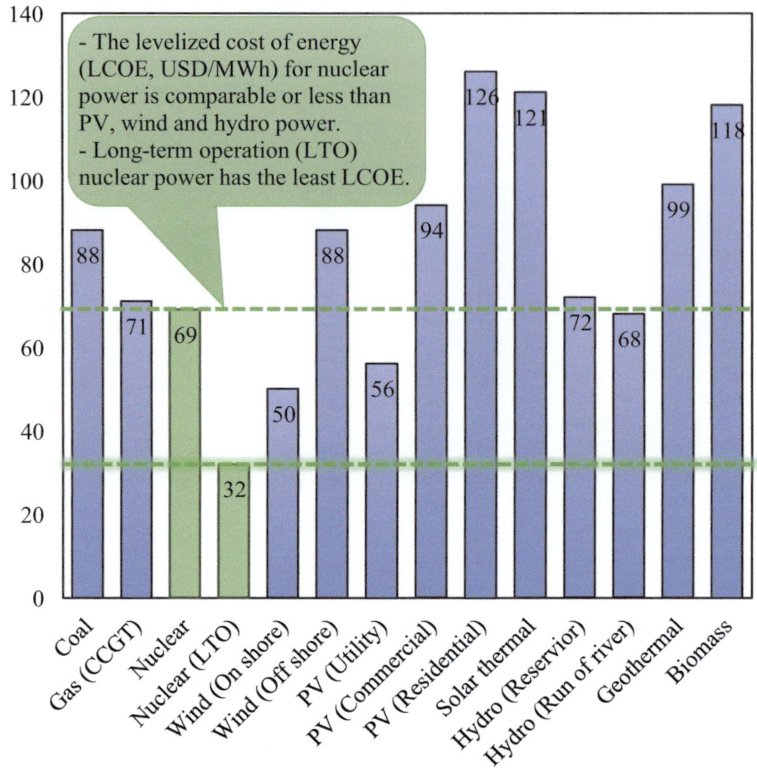

Fig. 5.3 LCOE (USD/MWh) of nuclear power generation compared with other technologies

extension enhances nuclear energy's competitiveness, making it the least costly option for low-carbon and overall power generation, as illustrated in Fig. 5.3.

While renewables prove highly competitive in many countries, they still face higher costs compared to fossil fuel and nuclear generation in specific regions, such as Japan, Korea, and Russia. In the United States, onshore wind and utility-scale solar PV are the most cost-effective options, followed by natural gas CCGTs, offshore wind, new nuclear build, and coal. In China and India, variable renewables like utility-scale solar PV and onshore wind offer the lowest expected generation costs, with nuclear energy also demonstrating competitiveness. The competitiveness of energy sources depends on national and local conditions, highlighting the evolving energy landscape.

Small Modular Reactors (SMR)

The research question surrounding Small Modular Reactors (SMR) centres on their potential as a greener and cheaper alternative to other renewable resources like solar and wind. SMRs offer a diverse range of nuclear technologies with lower costs, advanced safety features, and flexibility [5]. As Australia commits to building nuclear-powered submarines, the expertise gained could also support SMR deployment. These reactors are being commercialized to provide low-cost, 24/7 zero-emission heat and power, and their smaller size and inherent safety features enable faster deployment in various markets. Using SMRs as dispatchable power generators, there is no need for massive energy storage integration within grids so the cost of infrastructure will be reduced as well. SMRs can potentially replace aging power plants, support developing nations with limited infrastructure, and provide clean power for remote locations. Their design allows for operation in water-scarce conditions, offering a cost-effective and sustainable solution. By using serial production in factories and shipyards, SMRs reduce construction time and costs. With many baseload power stations set for retirement, SMRs are well-positioned to become a part of Australia's future energy mix. Overall, SMRs hold promise as a viable and efficient solution for zero-emission power supplies in various settings [6].

Micro-reactors, also known as very small SMRs (vSMRs), refer to nuclear power reactors generating less than 20 MW-thermal (MWt). The USA Office of Nuclear Energy defines the electric power level of micro-reactors as ranging between one and ten MWe. These self-contained, low-maintenance, and easily transportable units are designed to be 'plug-and-play' with self-regulating capabilities, reducing the need for specialized operators and preventing overheating. Micro-reactors serve a vital role in decentralized energy production, particularly in remote communities and industrial sites with low power demands, while also supporting applications such as district heating, water desalination, and hydrogen fuel production. Their mobility makes them a promising alternative to diesel power generators in off-grid areas and during emergencies, offering the potential for seamless integration with renewable energy sources within micro-grids [5].

Another advantage of SMRs lies in their reduced fuel demands. These innovative power plants differ significantly from conventional ones, as they may necessitate refueling only every 3 to 7 years, in stark contrast to the more frequent refueling intervals of 1 to 2 years seen in traditional plants. Remarkably, certain SMRs are engineered to run continuously for up to 30 years without the need for refueling, making them a compelling option for long-term, efficient energy generation [7]. Many SMRs are constructed or under development around the world.

Some SMRs are designed to be very versatile. For example, MoveluX reactor system by Toshiba is a multi-purpose energy source, serving various functions such as off-grid and microgrid electricity generation, high-temperature heat supply, and hydrogen production. Designed to produce 10 MW thermal, it offers flexibility for electricity production, allowing for output adjustments as needed. When integrated into a small or microgrid with load-following capabilities facilitated by a predictive control system, the MoveluX reactor system proves its adaptability as a multi-purpose energy source within such microgrids. Key provisions for resource optimization in the MoveluX reactor include a simplified plant design to minimize waste during operation and decommissioning, low maintenance requirements owing to no-moving parts components, resulting in reduced maintenance costs and labour needs, and a reduced emergency planning zone, which lessens the burden and costs associated with accident management and evacuation procedures [7].

As shown Fig. 5.4 some of SMRs are also specific to some regions with high heat demand, such as Russia. For example, ELENA nuclear thermoelectric plant (NTEP) stands as a remarkable innovation, representing a direct conversion water-cooled reactor that operates without the need for on-site refueling. This ingenious facility boasts the capability to provide 68 kW of electricity and a substantial 3.3 MW thermal heating capacity over an impressive 25-year period without requiring refueling. Remarkably, the unattended ELENA NTEP is engineered to serve as a heat source for towns with a population ranging from 1500 to 2000, primarily situated in remote areas where district heating is essential, thus addressing critical energy needs in such locations.

Detailed research should be conducted to evaluate more accurately lifetime emissions and LCOE of the energy delivery system using different technologies. For example, the following avenues should be studied:

- Although the LCOE of renewables are comparable with nuclear energy at the moment, however, other costs associated with the stability of the grid are not considered where there are high levels of PV and wind. Stabilizing the grid using energy storage and synchronous condensers adds additional costs to the system with intermittent renewables, which reduces the affordability of energy systems based on PV and wind and increases the viability of nuclear power. This is because nuclear power is dispatchable, and we don't need an energy system to stabilize it.
- The cost of investment using SMRs from their modularity and scalability perspectives [8] is much cheaper than large nuclear power plants, therefore, the long-term cost of the energy system seems to be cheaper using nuclear than intermittent energy systems such as PV and wind.
- The lifetime of SMR can be built and operated almost without service and refueling for their lifetime, which makes them very effective in terms of reliability and cost-effectiveness. Also, the lifetime of SMR is almost longer than PV and wind turbines.
- The reliability and safety of SMRs have increased significantly [9] so we can get more energy per investment cost from them at a lower risk to humans.

Another aspect that should be considered is that nuclear power plants face a range of potential hazards, including external events like structural pressure and missile impacts that can disrupt safety functions. Cooling and ventilation issues may result from external factors, leading to the loss of safety systems reliant on air cooling or exposure to toxic

Fig. 5.4 Small modular reactors' (SMRs) status around the world: current in-operation and under construction SMRs with nominal electrical power outputs, and the manufacturers involved in SMRs. © IAEA, 2022, https://aris.iaea.org/Publications/SMR_booklet_2022.pdf [7]

gases. Problems with the ultimate heat sink can cause partial or complete loss of secondary cooling and other safety measures dependent on water cooling. External power supply interruptions may result in a loss of off-site power, while events like external flooding and fires can disable safety functions and compromise the plant's structure. Additionally, external events like electrical disturbances can indirectly affect the plant by interfering with power supply or control signals to safety systems. These factors underline the complex challenges and safety considerations inherent in nuclear power plant operations, for which reliable solutions are available for many regions, but should be studied for new nuclear plants in new regions [3, 10–12].

Many qualitative and mixed methods investigations have addressed the psychological aspects of residing near nuclear power plants or experiencing nuclear accidents.

These studies have unveiled recurrent themes, which encompass a fundamental distrust in authorities, the significant role of one's personal history in shaping psychological experiences, and the influence of familiarity and attachment to a particular location in shaping individual perceptions. This research underscores the intricate interplay between psychological factors and proximity to nuclear facilities or exposure to nuclear accidents, shedding light on the multifaceted nature of these experiences. These aspects should also contribute to the new research on the expansion of nuclear [13, 14].

References

1. I. E. Agency, "Nuclear Power and Secure Energy Transitions From today's challenges to tomorrow's clean energy systems," INTERNATIONAL ENERGY Agency2022, Available: https://iea.blob.core.windows.net/assets/016228e1-42bd-4ca7-bad9-a227c4a40b04/NuclearPowerandSecureEnergyTransitions.pdf.
2. Iea. (2023). *CO2 emissions avoided by nuclear by country or region, 1971-2022.* Available: https://www.iea.org/energy-system/electricity/nuclear-power.
3. E. Choi, J.-G. Ha, D. Hahm, and M. K. Kim, "A review of multihazard risk assessment: Progress, potential, and challenges in the application to nuclear power plants," *International Journal of Disaster Risk Reduction,* vol. 53, p. 101933, 2021/02/01/ 2021.
4. Iea. (2023). *Nuclear Power - IEA.* Available: https://www.iea.org/energy-system/electricity/nuclear-power.
5. E. M. A. Hussein, "Emerging small modular nuclear power reactors: A critical review," *Physics Open,* vol. 5, p. 100038, 2020/12/01/ 2020.
6. M. Ho, E. Obbard, P. A. Burr, and G. Yeoh, "A review on the development of nuclear power reactors," *Energy Procedia,* vol. 160, pp. 459-466, 2019/02/01/ 2019.
7. IAEA, "Advances in Small Modular Reactor Technology Developments," 2022, Available: https://aris.iaea.org/Publications/SMR_booklet_2022.pdf.
8. P. A. Wrigley, P. Wood, S. O'Neill, R. Hall, and D. Robertson, "Off-site modular construction and design in nuclear power: A systematic literature review," *Progress in Nuclear Energy,* vol. 134, p. 103664, 2021/04/01/ 2021.
9. R. Bruce Hayes, "Nuclear energy myths versus facts support it's expanded use - a review," *Cleaner Energy Systems,* vol. 2, p. 100009, 2022/07/01/ 2022.

10. T. A. Kurniawan *et al.*, "Technological solutions for long-term storage of partially used nuclear waste: A critical review," *Annals of Nuclear Energy,* vol. 166, p. 108736, 2022/02/01/ 2022.
11. H. Lin *et al.*, "A review on the risk, prevention and control of cooling water intake blockage in coastal nuclear power plants," *Nuclear Engineering and Technology,* 2023/10/10/ 2023.
12. K. Ishiki, L. S. Cunningham, and B. D. Rogers, "Existing design approaches to nuclear power plants subject to tsunamis: A critical review," *Structures,* vol. 57, p. 105109, 2023/11/01/ 2023.
13. M. W. Edwards, R. D. Schweitzer, J. Shakespeare-Finch, A. Byrne, and K. Gordon-King, "Living with nuclear energy: A systematic review of the psychological consequences of nuclear power," *Energy Research & Social Science,* vol. 47, pp. 1-15, 2019/01/01/ 2019.
14. C. Greenhalgh and A. Azapagic, "Review of drivers and barriers for nuclear power in the UK," *Environmental Science & Policy,* vol. 12, no. 7, pp. 1052-1067, 2009/11/01/ 2009.

6 Real Cost of Electricity and Renewables

Understanding the real cost of electricity is fundamental to evaluating the sustainability and fairness of future energy systems. This chapter explores the intricate relationship between generation technologies, market structures, and policy frameworks in shaping electricity prices. It highlights the multiple drivers behind cost trends, including fuel volatility, infrastructure investment, technological progress, and the integration of renewable resources. Beyond direct production costs, the chapter examines how hidden factors such as system reliability, storage requirements, and environmental externalities redefine the notion of affordability. By critically assessing both conventional and renewable sources through an economic lens, the discussion provides insight into the true financial implications of transitioning toward clean energy and underscores the need for holistic approaches to ensure equitable access to electricity.

Electricity Price Trend

According to the Australian Energy Council, household electricity prices in Australia have doubled between 2007 and 2014. This increase has been well above the consumer price index (CPI) in Australia since 2007 onward, which shows that electricity prices have increased at much higher rates than inflation. Australian residential prices are at 19.21 US c/kWh, just around the OECD average, which is 19.8 US c/kWh [1]. A comparison of residential electricity prices per kWh in different countries is provided in Fig. 6.1 [2]. As seen, the average for Australian energy prices is just higher than the average within one standard deviation of the average. It is important to note that the fixed rate or connection fee per day for residential is increasing when the level of PV and battery adoption at the

Md. R. Islam et al., *Universal Access to Sustainable, Affordable, and Reliable Electricity*, Synthesis Lectures on Sustainable Development,
https://doi.org/10.1007/978-3-032-19517-3_6

household level is also increasing. This fact increases the amount of electricity price by the time where the self-energy consumption grows.

The increasing electricity prices show themselves in extreme events and price spikes in electricity markets as well. The rapidly changing energy landscape presents a critical need to address extreme events in electricity spot prices and ensure peak demand can be met during supply shortages. Policymakers should consider regulatory mechanisms to sustain system reliability, particularly in the face of price drops that may lead to reduced capacity stock [3]. Researchers should explore incorporating comprehensive factors for developing a better forecasting algorithm, as it can detect changes more timely for stabilizing the market and reducing shocks to the industry [4]. Improved electricity price forecasting offers valuable insights into end-customer clustering, aiding energy providers in identifying usage patterns and enabling personalized marketing strategies like new tariffs and incentives. This presents an opportunity for utilities and retail energy providers to update their load forecasting practices effectively.

Factors of Electricity Prices

The source of energy for generating electricity plays a critical role in determining electricity prices. For example, the electricity price dynamics across Europe in 2021 were influenced by various factors. Fossil fuels, primarily coal and natural gas, played a significant role in shaping electricity prices. Coal-based generation had a substantial impact in Poland, the Czech Republic, Bulgaria, and Germany, with coal determining prices over 70% of the time in the first three countries and approximately 48% in Germany [5, 6]. On the other hand, natural gas dominated price formation in Belgium, Spain, Italy, the Netherlands, and Great Britain.

Surprisingly, despite non-fossil power generation accounting for 66% of electricity generation in Europe that year, fossil fuels still set electricity prices for 58% of the hours in the EU-27, showcasing their enduring influence. Notably, some Nordic countries, such as Sweden and Norway, relied heavily on hydropower plants to set electricity prices throughout the year [5]. Additionally, cross-border electricity imports played a pivotal role in price determination, particularly in countries like Hungary, Croatia, and Lithuania, where over 50% of their annual demand was met through imports. Denmark's electricity prices were strongly influenced by neighboring countries like Norway, Sweden, and Germany, with Denmark being price-dependent on these nations for 57% of the time in 2021. Denmark's domestic electricity generation, primarily wind, solar PV, and CHP, which largely ran based on heat demand, accounted for only 43% of the year's marginal production [5, 7].

Another aspect that influences electricity prices is establishing electricity markets, which were mainly developed to reduce the price of electricity by including the component of competition among providers. Electricity market restructuring, a global

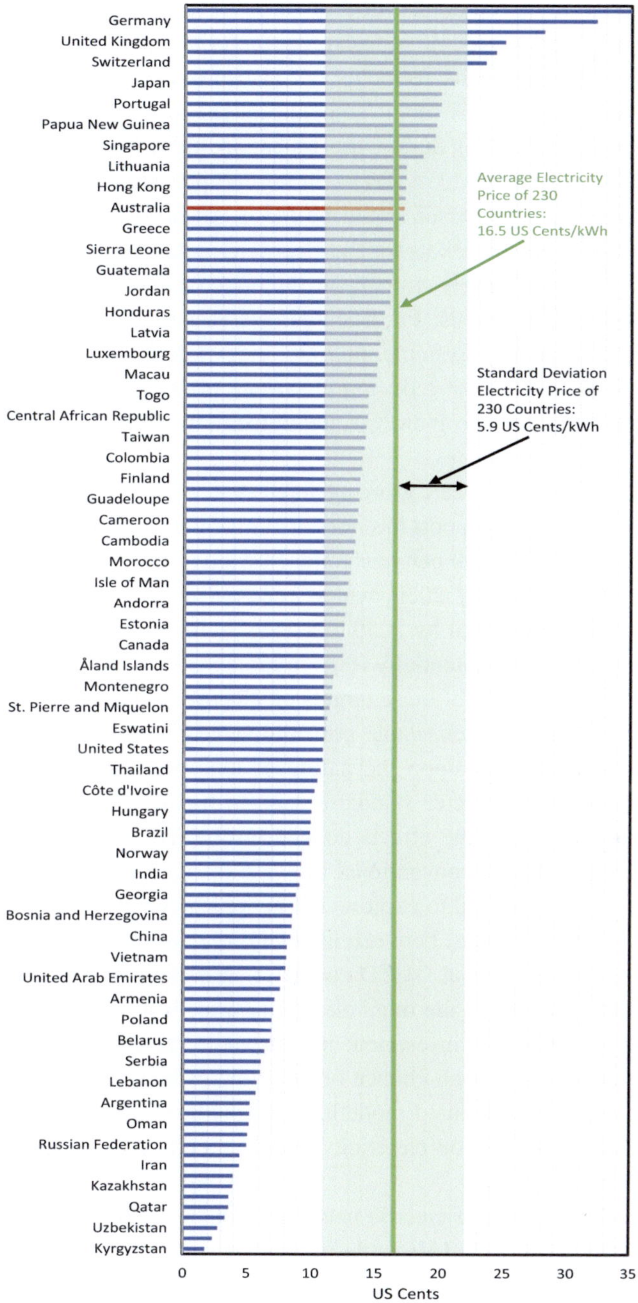

Fig. 6.1 Household electricity prices of selected countries out of the 230 countries studied

phenomenon, varies in its outcomes due to differing pre-restructuring conditions. For example, analyzing the impact of industry restructuring on retail electricity prices in Texas is not only relevant to its residents but also provides valuable insights for market reforms elsewhere. In Texas, retail competition led to an increase in average prices by $0.0112/kWh ($11.2/MWh) during the transition phase from 2001 to 2006, and a further rise of $0.0134/kWh ($13.4/MWh) during the period of unrestricted competition from 2007 to 2020 [8]. One reason could be the overhead of running an administrative section for running different markets for the electricity industry, that was not previously in place. Researchers should deeply investigate the effectiveness of electricity markets and their impact on electricity prices. Furthermore, the structure of power system management should be reviewed. For example, in 2006, the wholesale electricity market (WEM) was established with the hope of reducing electricity prices by increasing competition in the electricity generation sector. However, the outcome was that the electricity price was doubled.

In recent years, there has been growing interest in green targets, such as renewable portfolio standards, but their support has not always been unanimous. For example, renewables subsidies such as $500,000 per year for each wind turbine, bringing the total support to renewables to $22 billion by 2030, which goes out of Australia [9]. Considering the estimate of $8 trillion investment by 2050 to achieve zero-net emissions, it seems that this trend would continue, which potentially leads to higher electricity prices in Australia. One potential concern is the impact of these targets on electricity prices. One study delves into the long-term implications of renewable portfolio standards on electricity prices, finding that the effects are contingent upon the nature of long-term average costs. When average costs remain constant, stricter standards lead to higher electricity prices. However, when average costs increase, the effects depend on the rate at which green electricity's cost rises compared to that of conventional brown electricity [10]. A detailed analysis by researchers should be conducted to explore distinctions between long-term and short-term effects, as well as the differences between intensifying existing standards and introducing entirely new ones. A study about OECD countries shows that where the share of renewable energy resources is higher, the retail electricity price is higher as well [11]. However, some models suggest that overinvestment in renewable resources such as PV and wind turbines will result in a very high chance of very low or negative prices of electricity by 2030 [12, 13]. Therefore, a detailed modelling and analysis should be conducted to evaluate the impact of renewables on electricity prices, specifically in the context of carbon pricing [14].

Another reason is the cost of electric networks, which is about 50% of the customer's electricity bill in Australia [15, 16], while in many other countries with more dense networks, the factor is less than 20%. As seen in Fig. 6.2, the share of the grid in retail price is between 30 and 60%, showing a significant contribution to the price. The next major component in electricity price is retail and wholesale expenses, as discussed in this section.

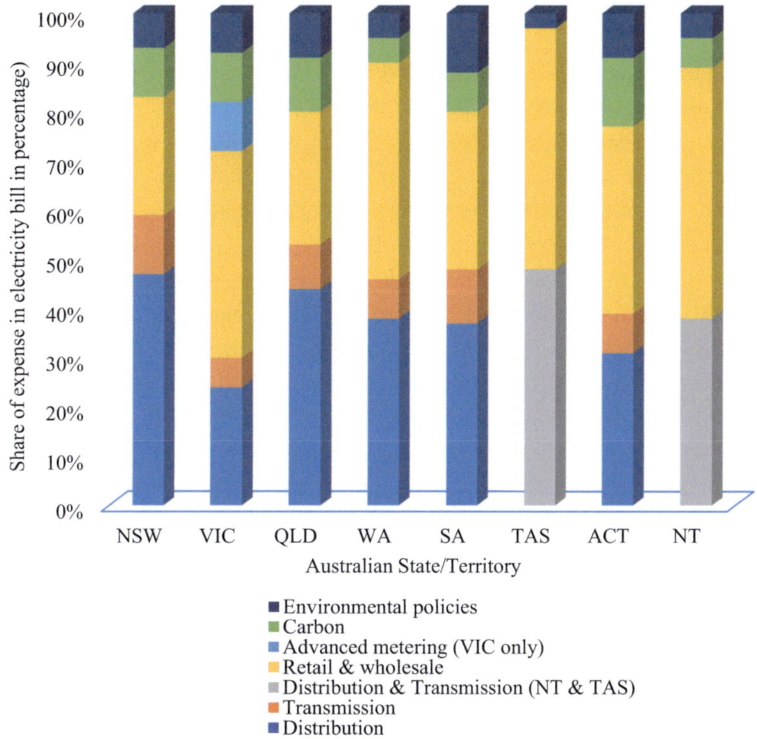

Fig. 6.2 The components of retail electricity bills in different States in Australia

Therefore, it is essential to research how to redesign transmission and distribution networks in Australia. The current practices are usually based on the experiences of countries with denser networks. In many areas in Australia, the use of microgrids and distributed generations in an islanded grid can be useful to reduce the excessive expansion of grids.

Further, some extreme events around the world would contribute to the increase of electricity prices, such as during the years 2020 to 2022, a period marked by Covid-19 lockdowns and the local conflict, which contributes to various shocks on electricity prices in some places, such as the UK [6]. Additionally, consumer demand shocks contributed to rising real electricity prices, particularly when Covid-19 restrictions were eased in the UK. Notably, the most significant contributor to the increase in real electricity prices was electricity-specific shocks. Moreover, there is compelling evidence suggesting that natural gas price shocks, consumer demand shocks, and electricity-specific shocks had adverse effects on real industrial production during this period [6].

Real Costs of Renewables

It is estimated that the cost of achieving zero net emissions is approximately $8 trillion by 2050 in Australia, in which the full figures of supporting elements are not modelled [17]. According to the Global Trends in Renewable Energy Investment Report in 2019 [18], the share of renewables in supplying energy worldwide increased from 5.2% in 2007 to 12.9% in 2018. During this decade, the investment in renewable energy capacity only (excluding large hydro), has been US$2.6 trillion [18]. How can we comprehensively assess the full economic implications and sustainability challenges of transitioning to 100% renewable energy supply, considering not only the investment in renewable technologies but also the costs associated with essential supporting elements such as energy storages, inertia providers, communication and security infrastructure, and the complete lifecycle of manufacturing renewable technologies including the cost of rare mineral mining, and recycling them? While solar and wind power are abundant and emit fewer greenhouse gases during operation compared to traditional fossil fuels, challenges arise when considering their intermittency. For instance, solar power generation is halted at night, necessitating a holistic assessment of costs over a full day to accurately compare with traditional sources like diesel. This highlights the importance of accounting for the variability in renewable energy generation when analysing their economic viability [19].

Supporting technologies play a crucial role in overcoming the intermittency of renewable resources. Energy storage systems, such as batteries, are essential for storing excess energy generated during peak production periods for use during times of low generation. Additionally, synchronous condensers enhance grid stability by providing reactive power support, mitigating voltage instability caused by fluctuations in renewable energy output [20]. These technological advancements are vital for ensuring the reliability and resilience of renewable energy systems, thus making them technically acceptable for widespread adoption. Therefore, when evaluating the true cost of renewables, the cost associated with supporting technologies also should be considered.

In addition to energy storage systems and grid stability technologies, the deployment of renewable energy also necessitates robust communication and security infrastructure, which can significantly impact the overall cost. Advanced communication networks are essential for monitoring and managing distributed renewable energy systems, optimizing energy flows, and coordinating grid operations in real time [21]. Moreover, cybersecurity measures are critical to safeguarding renewable energy infrastructure against potential cyber threats and ensuring the integrity and reliability of energy supply. The investment required for developing and maintaining these communication and security systems adds to the overall cost of renewable energy deployment. Therefore, it is essential to consider the impact of these supporting technologies on the real cost of renewables to accurately assess their economic viability and sustainability.

While renewable energy sources have lower operational emissions compared to fossil fuels, their total environmental impact must be assessed over their entire lifecycle

[22]. The manufacturing process of solar photovoltaic panels and wind turbines involves mining and processing raw materials, leading to greenhouse gas emissions, water usage, and soil pollution. Moreover, the disposal of renewable energy infrastructure at the end of its lifespan poses environmental challenges. Therefore, a comprehensive evaluation of the emissions and pollution associated with renewable technologies is essential to accurately gauge their environmental benefits. To fully understand the real cost of renewables, it is imperative to redefine the concept of renewable energy, considering the lifetime impacts and operational limitations of renewable technologies [23]. This entails accounting for the environmental and social costs associated with the entire lifecycle of renewable energy systems, including manufacturing, installation, operation, and decommissioning. By adopting a holistic approach that integrates economic, environmental, and social considerations, policymakers and stakeholders can develop more accurate assessments of the true costs and benefits of renewable energy deployment. Such an approach will facilitate informed decision-making and promote the transition towards a sustainable energy future.

References

1. C. Tran. (2016, APR 28 2016). *Worldwide electricity prices: How does Australia compare?* Available: https://www.energycouncil.com.au/analysis/worldwide-electricity-prices-how-does-australia-compare/.
2. (2023). *Worldwide Electricity Pricing | Energy Cost Per KWh in 230 Countries.* Available: https://www.cable.co.uk/energy/worldwide-pricing/.
3. N. Apergis, W.-F. Pan, J. Reade, and S. Wang, "Modelling Australian electricity prices using indicator saturation," *Energy Economics,* vol. 120, p. 106616, 2023/04/01/ 2023.
4. S. Gupta and M. Y. Bhat, "Electricity Price spirals now or later: An emerging market experience," *The Electricity Journal,* vol. 36, no. 5, p. 107272, 2023/06/01/ 2023.
5. B. Zakeri *et al.*, "The role of natural gas in setting electricity prices in Europe," *Energy Reports,* vol. 10, pp. 2778–2792, 2023/11/01/ 2023.
6. C. N. Ganepola, M. Shubita, and L. Lee, "The electric shock: Causes and consequences of electricity prices in the United Kingdom," *Energy Economics,* vol. 126, p. 107030, 2023/10/01/ 2023.
7. E. Hosius, J. V. Seebaß, B. Wacker, and J. C. Schlüter, "The impact of offshore wind energy on Northern European wholesale electricity prices," *Applied Energy,* vol. 341, p. 120910, 2023/07/01/ 2023.
8. J. Zarnikau, K. H. Cao, H. S. Qi, and C. K. Woo, "Has retail competition reduced residential electricity prices in Texas?," *Utilities Policy,* vol. 84, p. 101657, 2023/10/01/ 2023.
9. P. o. Australia. (2023). *Chapter 7 – Parliament of Australia.* Available: https://www.aph.gov.au/parliamentary_business/committees/senate/wind_turbines/wind_turbines/Final%20Report/c07.
10. K. Lee, "Renewable portfolio standards and electricity prices," *Energy Economics,* vol. 126, p. 106959, 2023/10/01/ 2023.
11. A. M. Oosthuizen, R. Inglesi-Lotz, and G. A. Thopil, "The relationship between renewable energy and retail electricity prices: Panel evidence from OECD countries," *Energy,* vol. 238, p. 121790, 2022/01/01/ 2022.

12. D. S. Mallapragada, C. Junge, C. Wang, H. Pfeifenberger, P. L. Joskow, and R. Schmalensee, "Electricity pricing challenges in future renewables-dominant power systems," *Energy Economics,* vol. 126, p. 106981, 2023/10/01/ 2023.
13. A. Biber, M. Felder, C. Wieland, and H. Spliethoff, "Negative price spiral caused by renewables? Electricity price prediction on the German market for 2030," *The Electricity Journal,* vol. 35, no. 8, p. 107188, 2022/10/01/ 2022.
14. K. Zhang, Y.-F. Yao, Q.-M. Liang, and G. Saren, "How should China prioritize the deregulation of electricity prices in the context of carbon pricing? A computable general equilibrium analysis," *Energy Economics,* vol. 96, p. 105187, 2021/04/01/ 2021.
15. A. Arefi, A. Abeygunawardana, and G. Ledwich, "A New Risk-Managed Planning of Electric Distribution Network Incorporating Customer Engagement and Temporary Solutions," *IEEE Transactions on Sustainable Energy,* vol. 7, no. 4, pp. 1646-1661, 2016.
16. E. N. Association. (2023). *ELECTRICITY PRICES AND NETWORK COSTS*. Available: https://www.energynetworks.com.au/assets/uploads/electricity-prices-and-network-costs_2.pdf.
17. A. Government, "Australia's Long-Term Emissions Reduction Plan: A whole-of-economy Plan to achieve net zero emissions by 2050," 2021, Available: https://www.dcceew.gov.au/sites/default/files/documents/australias-long-term-emissions-reduction-plan.pdf.
18. "Global Trends in Renewable Energy Investment," 30 July 2019 2019, Available: https://www.unep.org/resources/report/global-trends-renewable-energy-investment-2019.
19. I. Akhtar, S. Kirmani, and M. Jameel, "Reliability Assessment of Power System Considering the Impact of Renewable Energy Sources Integration Into Grid With Advanced Intelligent Strategies," *IEEE Access,* vol. 9, pp. 32485-32497, 2021.
20. A. Roux, B. Bekker, and A. Dalton, "Synchronous condensers as a viable inertia support mechanism on the future South African grid," *Energy for Sustainable Development,* vol. 69, pp. 192–201, 2022/08/01/ 2022.
21. B. Behi, A. Arefi, P. Jennings, A. Gorjy, and A. Pivrikas, "Advanced monitoring and control system for virtual power plants for enabling customer engagement and market participation," *Energies*, vol. 14, no. 4. https://doi.org/10.3390/en14041113.
22. W. N. Association, "Comparison of lifecycle greenhouse gas emissions of various electricity generation sources," 2011, Available: https://www.world-nuclear.org/uploadedfiles/org/wna/publications/working_group_reports/comparison_of_lifecycle.pdf.
23. A. Arefi, A. Abeygunawardana, and G. Ledwich, "A new risk-managed planning of electric distribution network incorporating customer engagement and temporary solutions," *IEEE Transactions on Sustainable Energy,* vol. PP, no. 99, pp. 1–1, 2016.

7 Universal and International Power Grid

The vision of interconnected electricity systems across national and regional boundaries presents both unprecedented opportunities and complex challenges for the future of energy. This chapter investigates the concept of universal and international power grids, emphasizing their potential to enhance energy security, facilitate large-scale renewable integration, and optimize resource utilization across diverse geographies. At the same time, it acknowledges the significant technical, economic, and political barriers that accompany such integration, including system interoperability, governance structures, and cross-border risk management. By analyzing the advantages, drawbacks, and decision-making frameworks associated with large-scale grid interconnection, the chapter situates universal and international grids as a critical yet demanding pathway toward achieving resilient and sustainable global electricity systems.

Universal Power Grid

As the penetration of renewables increases, different parts of the grid may require either AC (alternating current) or DC (direct current) grids to accommodate the characteristics of various renewable sources. Sometimes it is cheaper to have an AC grid and sometimes DC grids, or sometimes a grid at another frequency than the nominal 50 or 60 Hz [1]. The concept of a Universal Power Grid can be a new paradigm as a potential solution for reducing the cost of renewable integration, enabling seamless integration of grids with AC, DC, and other frequencies within power systems. In addition to this capability, Universal Power Grids are able to run at different frequencies without any changes in their systems. New research should be conducted on the Universal Grid concept, focusing on

Md. R. Islam et al., *Universal Access to Sustainable, Affordable, and Reliable Electricity*, Synthesis Lectures on Sustainable Development,
https://doi.org/10.1007/978-3-032-19517-3_7

its potential benefits, challenges, and implications for grid operation. Technical aspects, such as converter technologies, control strategies, and grid architecture, are necessary to implement a Universal Grid that accommodates both AC and DC flows. Additionally, the economic, regulatory, and policy considerations associated with the deployment of a Universal Power Grid (UPG) have to be discussed.

One of the effective approaches that brings together the benefits of AC and DC grids is hybrid AC/DC microgrids [2, 3]. Microgrids have received significant attention within the academic and different industry applications, serving as a promising solution for economic integration of distributed energy sources and energy storage systems. Traditionally, the predominant configuration has been the AC microgrid, but DC-based microgrids have been garnering increasing interest due to their distinct advantages, such as the absence of reactive power, elimination of synchronization requirements, and a growing array of DC-powered devices. Consequently, the emergence of hybrid AC/DC microgrids is growing as an effective strategy, offering a combination of the key benefits from both AC and DC microgrids [2, 4]. The overall concept of hybrid AC/DC microgrids is provided in Fig. 7.1.

Advantages and Challenges of Universal Power Grids

One of the inherent advantages of hybrid AC/DC microgrids is that AC- and DC-based devices can be linked to the grid more efficiently with minimal interface components, resulting in reduced conversion stages and consequently lower energy losses. This quality positions hybrid microgrids as an excellent choice for accommodating the growing number of DC-based units, including energy storage systems, EVs, PVs, fuel cells, laptops, and mobile phones, while still supporting AC-connected devices within the AC network. Moreover, these microgrids eliminate the need for synchronization between generation and storage units, as they can be directly connected to either the AC or DC network, simplifying control strategies. When it comes to voltage transformation, it's easily achievable on the AC side through transformers, whereas the DC side relies on DC-DC converters for conversion. Despite the higher initial cost associated with the inclusion of a power converter and communication network for connected devices in a hybrid microgrid compared to AC microgrids, this investment can be recouped more quickly as the number of connected devices increases, reducing the overall number of interface converters [5].

Nonetheless, the hybrid AC/DC microgrid approach is not without its share of shortcomings, which necessitate further investigation and detailed research in this domain. For example, the protection of such hybrid networks poses a considerable challenge, particularly in the context of DC-based networks. While protection devices for AC systems have undergone extensive research and development due to their prevalent use in the current utility grid, DC protection devices remain relatively uncharted territory [6]. Furthermore, fault detection in AC networks benefits from the simplicity afforded by zero-crossings

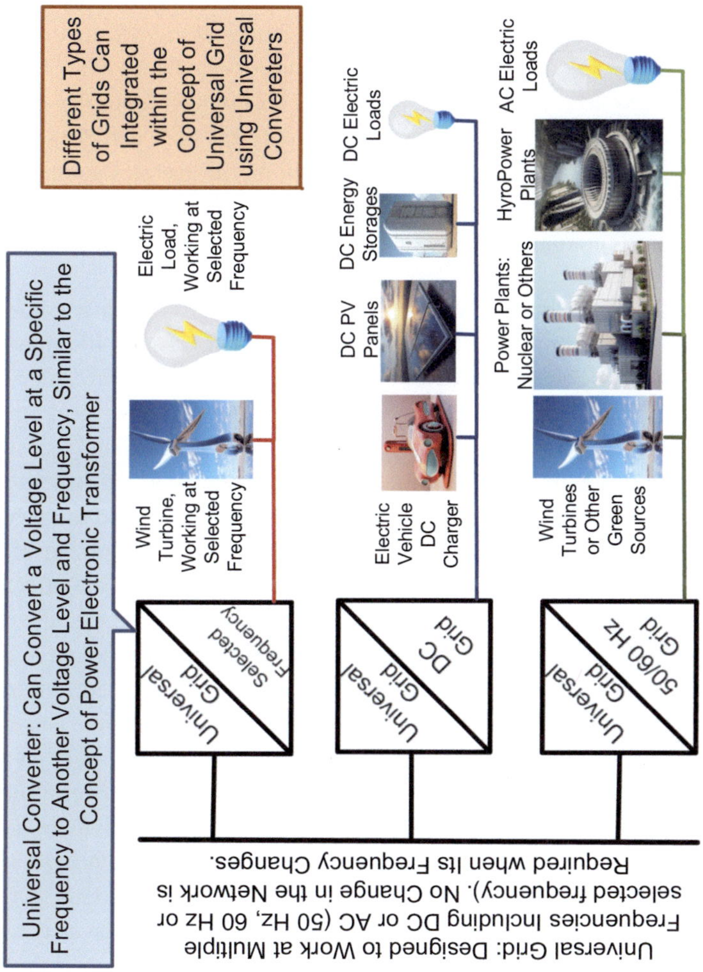

Fig. 7.1 The concept of universal power grids

of the current, a method that cannot be seamlessly transposed to DC grids [7]. Another challenge would be associated with reliability as hybrid microgrids tend to exhibit a lower degree of reliability when compared to their AC counterparts. This reduction in reliability can be attributed to the introduction of an interface power converter within the distribution network to facilitate the connection between AC and DC links. However, these networks experience an improved reliability for the connected devices, given the reduction in the number of converter stages [8]. In addition, the issue of control complexity looms large over hybrid microgrids. Managing these systems is inherently more intricate than their AC or DC counterparts due to the need for simultaneous control of devices connected to both AC and DC networks, as well as the interface power converter bridging the two [9, 10]. Consequently, ensuring a stable and reliable power supply for both networks becomes a challenging task, particularly in scenarios where the microgrid operates autonomously or in islanded mode or in the case that there are very high levels of intermittent renewable generation within hybrid microgrids [11]. Figure 7.2 shows the advantages and challenges of hybrid AC/DC microgrids.

Furthermore, volume considerations are significant, especially when it comes to the interface devices like transformers and power converters that connect the microgrid to the primary power network and to the devices. For example, converter-based devices tend to occupy less space than those reliant on transformers. Also, it is critical to develop transformer-less converters where possible in as many cases as. Additionally, cost considerations are integral to this hybrid model, which includes all the interface devices required for establishing the AC and DC networks within the microgrid. It is noted that when these stages are converter-based, the cost tends to rise due to the increased number of components involved. Conversely, power transformers are lauded for their cost-effectiveness, attributed to their ease of construction and minimal need for control or specialized maintenance [12]. Maintenance is another important aspect, where the comparison is based on the number of power conversion stages. Systems with more electronic converters generally demand more maintenance. Scalability is a pivotal factor, assessing the capability for voltage level variation dependent on the load. Power converters can be easily parallelized or serialized to match the instantaneous power flow, thereby boosting overall efficiency and reducing component wear and tear. In contrast, transformers necessitate additional components, like tap changers, to offer similar flexibility. Lastly, modularity is a critical parameter, measuring the capacity to augment the rated power of conversion stages. This variable highlights that the addition of new branches to the original power converter in parallel or series can enhance their rated power, whereas transformers require the addition of new units in parallel to operate at higher rated power levels, making modularity a more complex task [13].

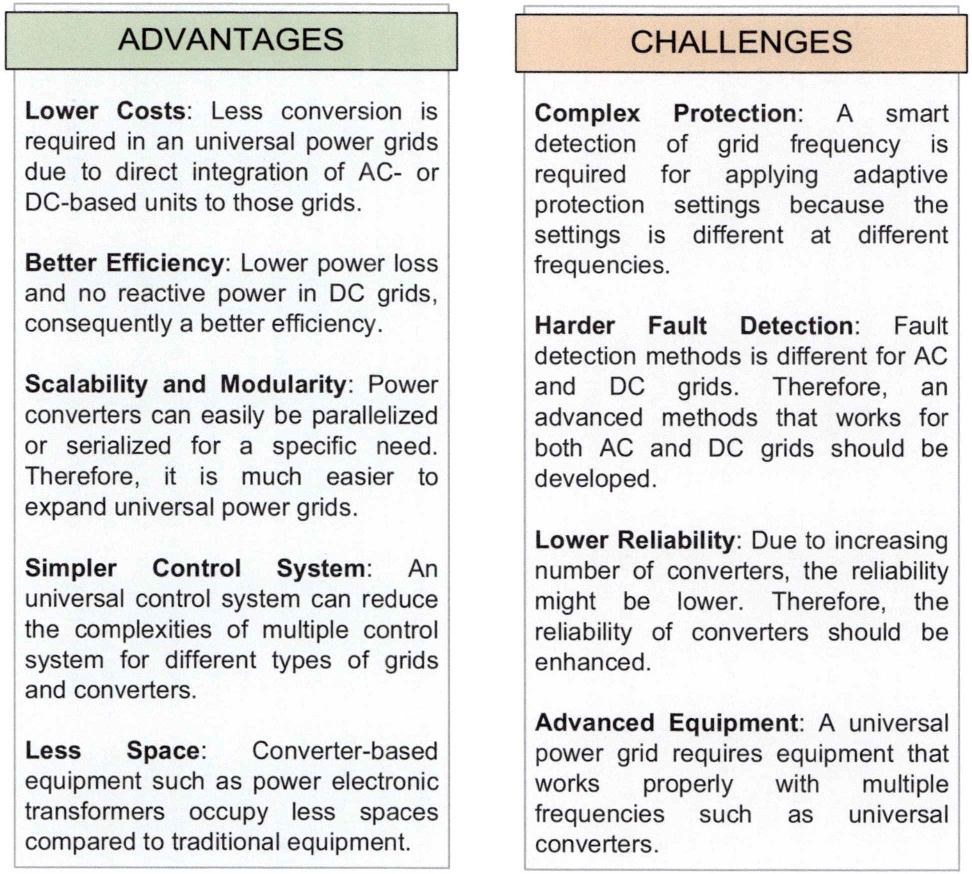

Fig. 7.2 The advantages and challenges of universal power grids

Fault Tolerant and Versatile Universal Power Grids

In ensuring the continuity of microgrid services, a fault tolerance control system is essential, which is accomplished through a series of techniques. First and foremost, there's the critical process of fault detection, which is the identification of any abnormalities or faults within the microgrid [14]. Following this, the recovery system comes into play, replacing the compromised state with a secure one. Within this phase, two primary mechanisms play a pivotal role. The first is fault management, involving the elimination of the fault in every erroneous state of microgrids, which is achieved through methods such as Rollback, Rollforward, and compensation [14]. The second mechanism, fault handling, encompasses several essential steps, including diagnosis to pinpoint the cause, type, and location of the fault, isolation to segregate faulty components or disable the affected zone, reconfiguration to assign new objectives to components functioning normally, and reset to verify, update, and record the microgrid states [15, 16]. These techniques collectively contribute to the robustness and uninterrupted operation of the system or microgrid. The protection system of a universal AC/DC or multifrequency microgrid needs to be studied in detail, and robust approaches and mechanisms should be evaluated and developed. The advanced fault tolerant control and protection system concept for an AC/DC microgrid are provided in Fig. 7.3.

In addition, the findings of studies unequivocally suggest that there is a small-signal stability issues in AC/DC microgrids, which arise primarily from the control systems governing the integrated converters, including voltage source inverters, the phase-locked loop, coordination and communication between converters. In cases where the parameters of these converter control systems are not appropriately configured, the converters may introduce negative resistance or create open-loop modal coupling, ultimately posing a threat to small-signal stability [17]. Furthermore, a standalone microgrid or a grid-connected microgrid through a converter could face small-signal stability issues under conditions of weak grid connection [18]. Additionally, the multi-terminal DC microgrids may experience instability when employing DC voltage droop control mechanisms. These challenges underscore the importance of precise control parameter settings in maintaining the stability of AC/DC microgrids, so a deep research should be conducted to provide a very stable and fault-tolerant control system for such microgrids [19].

Recently, there is a growing trend in versatile designs of equipment and devices to be able to connect directly to both DC and AC grids. For example, many household appliances such as TVs, washing machines, dishwashers, refrigerators, and microwaves can work with AC or DC supply without any additional interface or converter. Also, there are efforts to develop a unified control system for converters in AC/DC grids to avoid any change or replacement when those converters are being used in AC or DC grids [20]. Another universal design is associated with the development of power electronics transformers that help with the change of voltage and frequency from any level to any level [21], which enhances the concept of universal power grids, introduced here. At the larger

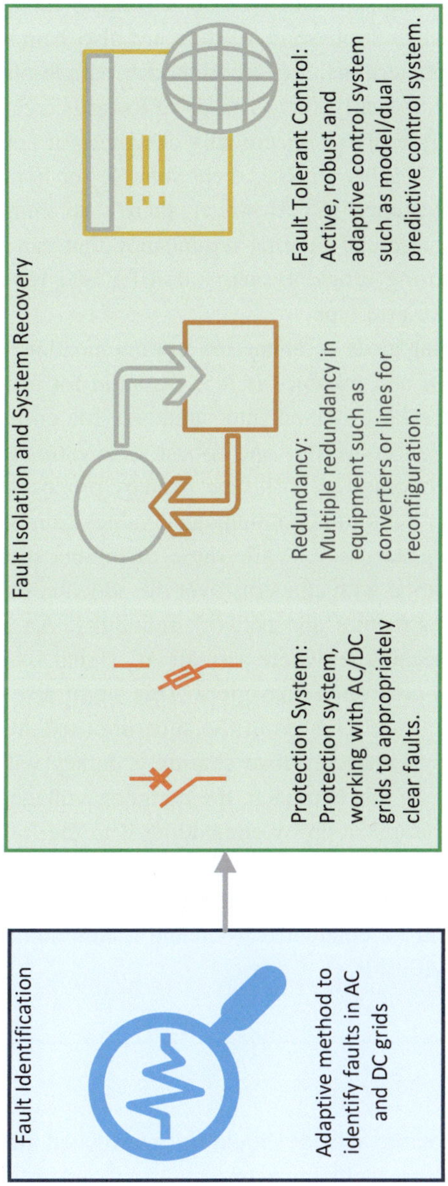

Fig. 7.3 The advanced fault tolerant control and protection system concept for a universal power grid

scale within grids, hybrid converters, equipped with DC and AC terminals, are becoming increasingly prevalent, particularly in the context of hybrid solar converters that offer an additional DC terminal for battery storage connection. Multiport converters, a more general category, are emerging, equipped with DC-link terminals for DC grid connections alongside AC grids. Universal multiphase converters are also being explored, featuring a high number of input/output terminals to accommodate single-phase AC, multiphase AC, and DC applications. Additionally, there is a drive towards creating power electronics solutions known as "energy routers," specifically designed for controlling power flow between residential DC and AC grids, making them suitable for both energy distribution systems and residential applications [22]. However, there's an imminent research need for universal converters with minimal internal redundancy that can seamlessly adapt to both DC and AC grids, ensuring efficient energy transfer and required reliability and controllability regardless of the grid type.

Another aspect of universal grids is about the communication between converters, which should be cost-effective and reliable, as it is essential for accurate control of the whole grid. Numerous studies have delved into strategies for ensuring precise current sharing among droop-controlled converters and robust state estimation through the utilization of communication networks [23, 24]. One strategy proposes adjusting reference signals for primary controllers based on communicated values from neighboring converters to enhance current sharing accuracy. While these communication-based approaches offer improved control, they come with concerns over the added costs associated with the required communication infrastructure and network reliability. An alternative approach involves the signal injection technique, where a small AC signal is superimposed on the DC grid voltage by all droop-controlled converters. This small active or reactive power flow, resulting from phase angle differences in the superimposed AC signals, is used to adjust output voltages, thereby enhancing load sharing accuracy without the need for a communication network [25]. In this approach, the common voltage information is estimated without the use of communication by embedding it in the frequency of the small AC signal, effectively making the signal frequency a global parameter in the DC grid, enabling the estimation of voltage magnitude in a communication-free manner. Therefore, extensive research should be conducted to enable a communication-less converter control system within universal grids.

International Grid

The costs and benefits of international grids should be investigated more thoroughly, especially from different perspectives, including economic, legal and political, environmental, and social.

Grid interconnections present a spectrum of economic and financial implications, encompassing both direct and indirect considerations. Directly, participating nations can reap economic benefits through the avoidance of costs associated with electricity generation, such as fuel purchase or production, capital outlays for generation facilities, operating expenses, and the construction and operation of transmission facilities. Countries engaged in interconnections may also derive direct financial gains through income from power sales, particularly advantageous for developing economies reliant on stable foreign currencies. Conversely, direct incurred costs involve expenses related to exporting electricity, operating generation facilities, building and maintaining the interconnection infrastructure, and purchasing power [26–28].

Indirectly, interconnections have the potential to stimulate economies by creating jobs during facility construction and operation. Additional indirect economic benefits may manifest in improved power supplies fostering local industry development, enhancements in education and healthcare, and a "re-spending" effect, whereby reduced electricity prices leave households with more disposable income for other consumption, savings, or investment. The interconnection structure, depending on the selling institution, may further stimulate power generation markets, contributing to decreased electricity prices. Effective pricing arrangements are essential, delineating the terms for electricity and electric system services exchanged through the interconnection. Pricing mechanisms, based on production costs, avoided costs, or negotiated agreements, including market-based options, become vital in fostering fair competition when sufficient buyers and sellers are present [29, 30].

International electrical grid interconnections present a dual-edged environmental scenario, offering a spectrum of benefits and impacts. On the positive side, such interconnections can yield advantages like diminished emissions of air pollutants on various scales, reduced water and solid waste pollution, minimized land-use effects, lowered impacts on biodiversity and wildlife, and enhanced human health outcomes. These benefits stem from the interconnection's influence on electricity generation and fuel usage across participating nations. However, it's crucial to acknowledge that net environmental impacts, both positive and negative, may arise in each category due to the interconnection. Moreover, the net environmental effects may vary across locations, necessitating a comprehensive assessment to ensure the inclusive consideration of all environmental costs and benefits [31, 32].

International grid interconnection initiatives present a nuanced landscape of social benefits and costs for diverse groups within participating nations. Anticipated advantages encompass improved power quality, heightened reliability, and broader electricity access for communities. If energy availability and reliability requirements of connected energy systems across countries are different, the marginal cost of reliability is an effective approach to set the required reliability expectations. In this approach, after determining the cost of marginal levels of reliability, each country should pay the cost associated with the level of reliability provided by another country. Additionally, there's the potential for income generation through power exports, specifically if directed towards social development objectives, and the fostering of cooperation between countries. However, potential social costs and liabilities include the risk of physically separating local groups from

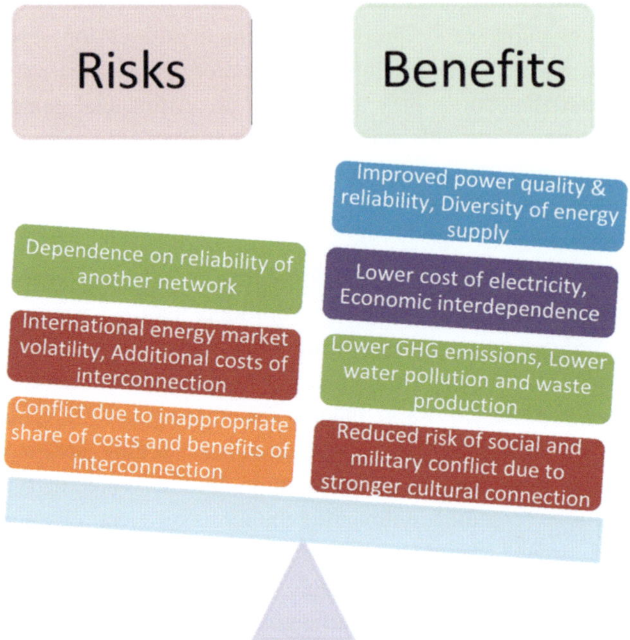

Fig. 7.4 Benefits and risks associated with international grids

their regular resources, introducing undesired external social influences during infrastructure construction, the social impacts associated with export power plant development, and a diminished incentive in power-importing nations to utilize local resources, thereby increasing susceptibility to supply disruptions. A comprehensive evaluation is imperative to weigh these social aspects and inform decision-making [33, 34]. The benefits and risks associated with international grids are summarized in Fig. 7.4.

International electric grid interconnections have the potential to yield political benefits for interconnected countries, fostering experience, political collaboration, conflict avoidance, and internal political stability. However, concerns may arise regarding perceived political and economic leverage, entanglement in internal affairs, and substantial political costs for power line protection. Establishing political agreements is crucial, encompassing resource sharing, project progression, payment structures for contractors, equitable distribution of benefits and costs, operational and security protocols for the interconnection infrastructure, governance agreements for the interconnection operator, and information sharing essential for planning, operation, and protection of the interconnection. These agreements, both between and within nations, form the foundation for the legal frameworks guiding international grid interconnections [35].

International electricity grid interconnections, barring their simplest forms, entail intricate legal processes involving various national, sub-national, and international entities to facilitate agreements for the planning, construction, and operation of cross-border power lines. The establishment of binding legal agreements necessitates transparency and enforceability, demanding national legal capabilities for drafting, reviewing, and adjudicating contract issues. Essential legal frameworks include power purchase and pricing agreements, agreements governing power line sitting and related infrastructure, operational agreements (including authorities), security agreements, environmental performance agreements, liability agreements for power line failure, and agreements governing the equitable selection of contractors for building, financing, and operating interconnection infrastructure. These legal instruments form the backbone of the complex landscape of international electricity grid interconnections [36, 37]. As discussed, the risks and benefits of international grids can be studied in five categories, including economic, environmental, social and cultural, legal, and political categories.

Decision Making for International Grids

For establishing an international grid, several criteria should be investigated to ensure secure and cost-effective power exchange between involved nations. Some of these criteria are as follows [38–42]:

- Ensure equitable distribution of economic, social, and other benefits and costs among nations and stakeholders.
- Specify direct and avoided costs accurately within comprehensive long-term power system planning.
- Emphasize transparency in all negotiations, providing stakeholders access to relevant materials.
- Involve all potentially affected parties early and solicit input throughout the project.
- Establish clear protocols for collecting and distributing quantitative data for project design and cost estimation.
- Assign clear legal and administrative authorities over all aspects of interconnection design, construction, and operation.
- Collaborate with international institutions, including financial ones, to facilitate political agreement and meaningful contributions from all groups.
- Preferably locate new power lines in existing transmission or transport corridors.
- Implement capacity building for social stakeholders to participate meaningfully in interconnection decisions.
- Consider energy supply indicators: total primary energy, the fraction of primary energy as imports, diversification indices, and stocks as a fraction of imports.
- Evaluate economic indicators: total energy system internal costs, total fuel costs, import fuel costs, and economic impact of fuel price increase.

- Assess technological indicators: diversification indices for key industries, diversity of r&d spending, reliance on proven technologies, and technological adaptability.
- Examine environmental indicators: GHG emissions, acid gas emissions, local air pollutants, other air and water pollutants, solid wastes, and nuclear waste.
- Consider ecosystem and aesthetic impacts, exposure to environmental risk, and social and cultural exposure to risk of conflict.
- Evaluate military/security exposure to risks and relative spending on energy-related security arrangements.

The decision-making criteria for establishing an international grid are illustrated in Fig. 7.5.

International Interconnection Cases

In Australia, the Australia-Asia Power Link (AAPowerLink) aims to harness and store solar energy from the Northern Territory of Australia, known for its abundant sunlight. This renewable energy will be transmitted to Darwin and Singapore 24/7 through a high voltage direct current (HVDC) transmission system. The AAPowerLink is set to become the world's largest solar farm and battery storage facility, covering a vast area and spanning a 5,000 km transmission network. Once operational, it will supply up to 15% of Singapore's total electricity needs, providing reliable and competitively priced renewable electricity. There has been some debate about the viability of this project [43].

High-voltage cable interconnectors are also connecting Europe's national power grids, which form the backbone of the integrated European electricity market, fostering the free flow of electricity across the continent [27]. Operated by transmission system operators, over 400 interconnectors link almost 600 million citizens, making Europe's electricity system the world's largest interconnected grid. Essential for achieving net-zero goals efficiently, interconnectors aid renewable integration, enhance power system efficiency, and reduce reliance on fossil fuels. Doubling Europe's interconnection capacity in the next decade is imperative for meeting energy and climate targets, as current expansion plans fall short. The lengthy nine-year average project timeframe necessitates immediate action to avoid hindering wind and solar power expansion, jeopardizing security of supply, and missing cost-saving opportunities. Multiple studies, including the Paris Agreement Compatible scenario, emphasize the critical role of interconnectors in realizing cost, flexibility, and reliability benefits [44]. There are many international interconnections around the world. The overall overview of international connections is illustrated in Fig. 7.6 [45].

In addition, an upcoming international link within the association of southeast Asian nations (ASEAN) involves establishing a 500 kV high-voltage direct current (HVDC) connection between the Philippine national grid and the Sabah power grid. This initiative aligns with the ASEAN economic community (AEC) objective, acknowledging climate

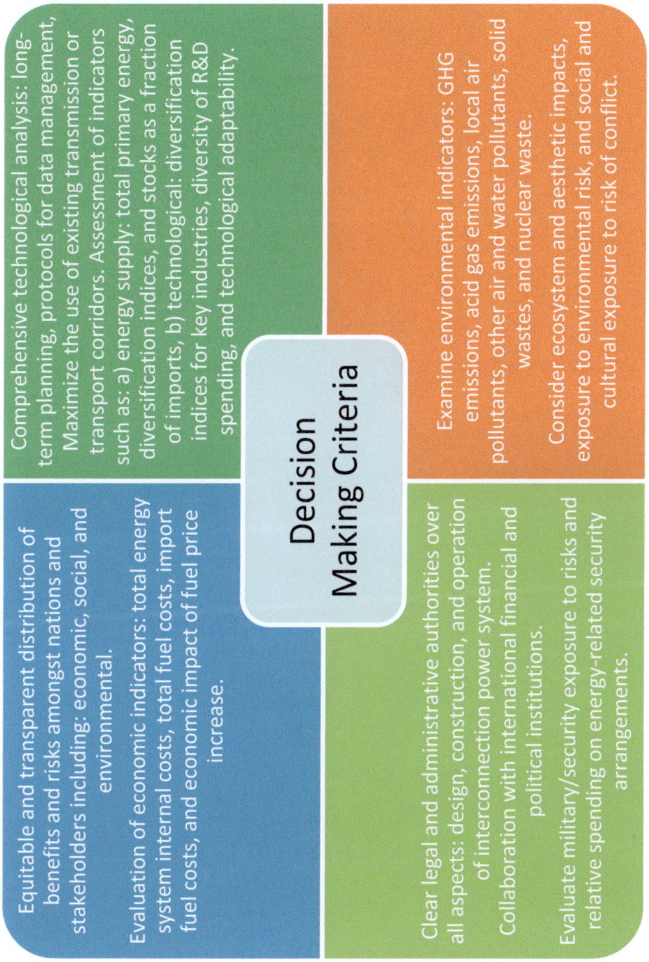

Fig. 7.5 The decision-making criteria for establishing an international grid

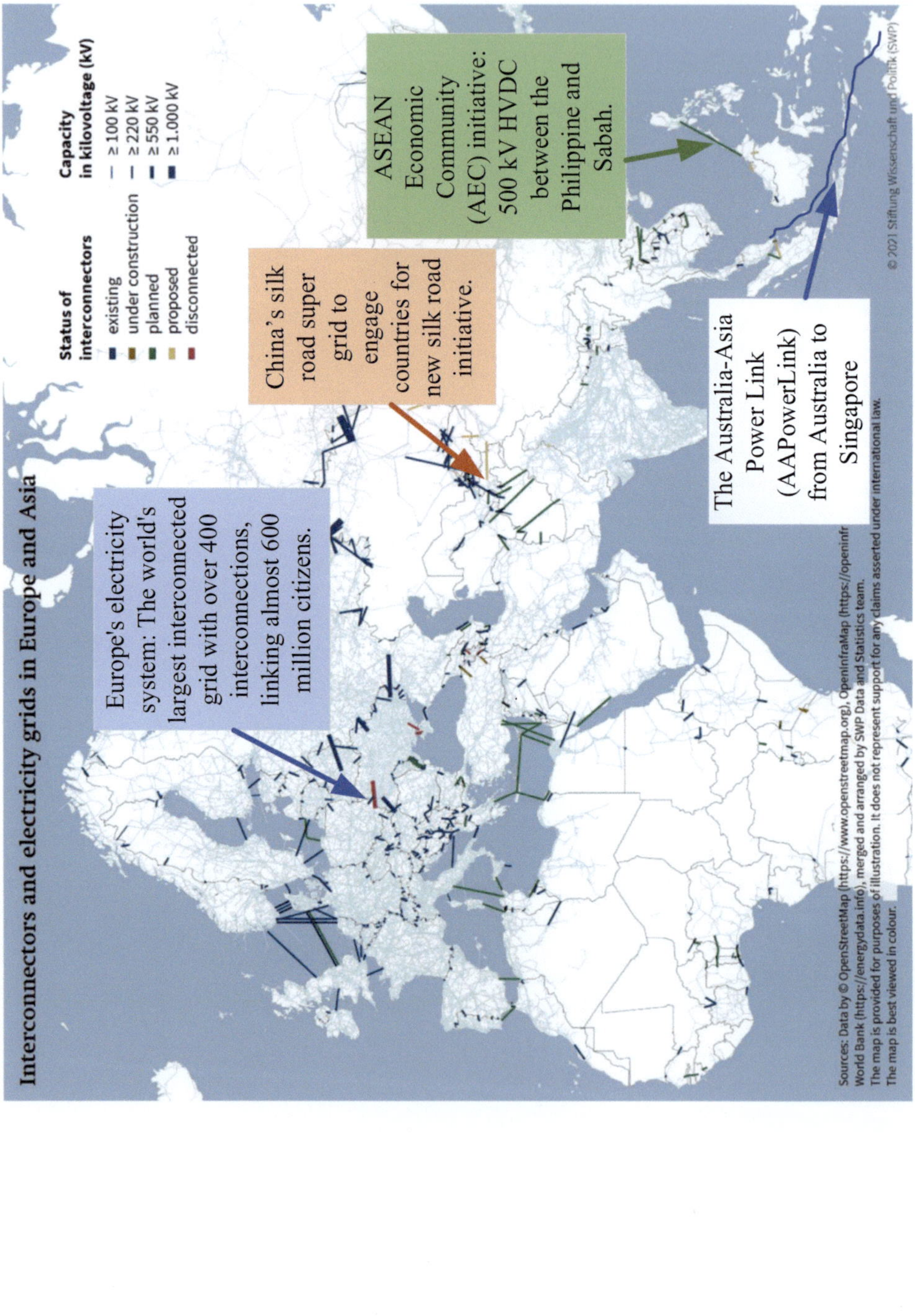

Fig. 7.6 The overall overview of international connections [45]

change's adverse effects. Committed to greener energy, the AEC aims to incorporate more renewables alongside fossil fuels, introducing complexity to the system's components and controls. The ASEAN interconnection master plan study (AIMS) II affirms the technical and economic feasibility of the proposed interconnection in the region, albeit without addressing small-signal stability and the impact of renewable energy sources in its network analyses [31].

China's engagement with neighboring national power grids can also be categorized into two phases. The initial phase, spanning from the 1990s to 2015, focused on cross-border line construction and power exchange to enhance collaboration with neighboring countries. The subsequent phase, initiated after 2015, centers on constructing an energy internet to achieve a low-carbon power system. This involves transnational interconnection, facilitating long-distance transmission of renewable power. Notably, the silk road super grid (SRSG) proposed by the international green technologies and investment centre in Kazakhstan explores the potential for trading affordable renewable energy along the new silk road. A feasibility study on power system interconnection between Kazakhstan, China, and South Korea demonstrates viability due to their complementary characteristics in renewable energy production, consumption patterns, and shared climate change mitigation targets. With a focus on load concentration in the east and abundant renewable energy sources and coal mines in remote areas, China has successfully implemented ultrahigh voltage direct current (UHVDC) projects, accumulating significant operational experience in long-distance electricity transmission. Studies have assessed transmission scenarios, considering costs and transmission prices based on economic and technical considerations [39].

Comprehensive Multicriteria Assessment Framework for New Projects in Energy Systems

There are several projects in the power system industry that are being identified, designed, engineered, and implemented. Some of those are very large-scale, such as EnergyConnect in Australia. It is very critical to develop a comprehensive framework for understanding the costs and benefits of such projects from different perspective, including financial, environmental, security, health, legal, etc. [46, 47]. For example, EnergyConnect is an important energy infrastructure project that aims to lower power bills, create jobs, and promote economic growth, primarily in regional NSW. It involves the construction of a new 900 km of 330 kV and 500 kV above-ground transmission lines, with approximately 800 MW transfer capacity, connecting power grids across three Australian states, which will increase competition in the wholesale electricity market. The project claims to bring several benefits, including $180 million in annual savings for NSW households, the creation of 1500 new jobs, and $4 billion in economic benefits, especially in regional areas [48]. Considering the major integration of intermittent renewable energies like solar and wind in the Australian energy landscape, how does the EnergyConnect project enhance security and reliability in the electricity supply, and whether this project is an optimum solution for increasing the reliability and security improvement of the grid?

In the evaluation of power system projects, economic and financial criteria play a crucial role in determining the feasibility and long-term viability of such endeavors. A summary of the proposed criteria for new projects in new energy systems is provided in Fig. 7.7. Detailed research projects must be conducted to develop the relevant parameters and performance indicators for each criterion.

The Levelized Cost of Electricity (LCOE) provides a vital metric for assessing the economic efficiency of a project by considering the total cost of electricity production over its lifespan. The initial investment and operational costs are fundamental factors influencing project viability, as they impact the overall financial performance. Evaluating the project's lifetime ensures its longevity and optimal return on investment. Accounting for lifecycle emissions and associated costs is essential for a comprehensive analysis, promoting

Fig. 7.7 A summary of the proposed criteria for new projects in new energy systems

environmentally responsible projects. Moreover, considering the benefits to the electricity market, such as increased reliability, reduced costs, and potential market competitiveness, further underscores the importance of robust economic and financial evaluation in shaping sustainable and successful power system projects [49–51].

In addition to the financial criteria for power system projects, environmental criteria play a pivotal role in evaluating power system projects, reflecting a growing awareness of the profound impact such endeavors can have on our ecosystems. Assessing air pollution ensures that the chosen energy sources don't compromise air quality, safeguarding human health and mitigating climate change. Similarly, scrutinizing water pollution protects water bodies, crucial for both human consumption and biodiversity. Land pollution considerations are essential to prevent soil degradation and maintain ecological balance. Evaluating human and cultural impact emphasizes the importance of respecting communities and preserving cultural heritage during project implementation. Lastly, assessing the impact on flora and fauna ensures that biodiversity is protected, contributing to the overall sustainability of power system projects. A comprehensive evaluation incorporating these environmental criteria is imperative to promote responsible and sustainable energy development [52–56].

Reliability and security criteria are paramount in the evaluation of power system projects, with a direct impact on the overall stability and sustainability of energy infrastructure. Integrating renewable sources, while crucial for environmental reasons, poses challenges to the traditional grid's reliability due to their intermittent nature. Evaluating the improvement of reliability and security with increased renewable integration is essential to ensure a seamless transition. Additionally, the cost of necessary network upgrades to maintain system stability during project implementation must be considered. As power systems move towards zero-net emissions, emphasizing reliability and security becomes even more critical. This transition involves not only technological advancements but also comprehensive planning to address potential vulnerabilities and risks, reinforcing the significance of these criteria in shaping resilient and environmentally sustainable power systems [57, 58].

Human resources development is integral to the evaluation of power system projects as it extends beyond the technical aspects, encompassing social and cultural dimensions. These projects have the potential to stimulate job creation, providing employment opportunities and contributing to economic growth. Moreover, emphasizing education and gender equality within the workforce ensures a diverse and skilled talent pool, fostering innovation and sustainability. Skills development is crucial to adapt to evolving technologies and industry requirements, enhancing the overall efficiency of power systems. The inclusion of different cultures promotes a rich and collaborative work environment, fostering creativity and understanding. By prioritizing human resources development, power system projects not only advance technological goals but also contribute to social progress, creating a workforce that is not only competent but diverse, inclusive, and equipped for the challenges of the future [59].

Further, incorporating technical criteria is indispensable in the evaluation of power system projects, ensuring the reliable and efficient operation of the energy infrastructure. Adhering to voltage and current limits is crucial to maintain system stability and prevent equipment damage. Frequency stability, another vital aspect, ensures synchronization within the power grid, preventing disruptions and enhancing reliability. Assessing the impacts of market rules is essential, as they influence the project's economic viability and integration into the broader energy landscape. Technical criteria provide the foundation for robust and resilient power systems, promoting seamless electricity generation, transmission, and distribution. By prioritizing these technical considerations, power system projects can navigate challenges, optimize performance, and contribute to a reliable and secure energy future [51, 60–62].

References

1. J. Ruddy, R. Meere, and T. O'Donnell, "Low Frequency AC transmission for offshore wind power: A review," *Renewable and Sustainable Energy Reviews,* vol. 56, pp. 75–86, 2016/04/01/ 2016.
2. E. Unamuno and J. A. Barrena, "Hybrid ac/dc microgrids—Part I: Review and classification of topologies," *Renewable and Sustainable Energy Reviews,* vol. 52, pp. 1251–1259, 2015/12/01/ 2015.
3. F. Nejabatkhah and Y. W. Li, "Overview of Power Management Strategies of Hybrid AC/DC Microgrid," *IEEE Transactions on Power Electronics,* vol. 30, no. 12, pp. 7072-7089, 2015.
4. B. Wei, X. Han, P. Wang, H. Yu, W. Li, and L. Guo, "Temporally Coordinated Energy Management for AC/DC Hybrid Microgrid Considering Dynamic Conversion Efficiency of Bidirectional AC/DC Converter," *IEEE Access,* vol. 8, pp. 70878-70889, 2020.
5. E. Unamuno and J. A. Barrena, "Hybrid ac/dc microgrids—Part II: Review and classification of control strategies," *Renewable and Sustainable Energy Reviews,* vol. 52, pp. 1123–1134, 2015/12/01/ 2015.
6. S. Mirsaeidi, X. Dong, and D. M. Said, "Towards hybrid AC/DC microgrids: Critical analysis and classification of protection strategies," *Renewable and Sustainable Energy Reviews,* vol. 90, pp. 97–103, 2018/07/01/ 2018.
7. B. Modu, M. P. Abdullah, M. A. Sanusi, and M. F. Hamza, "DC-based microgrid: Topologies, control schemes, and implementations," *Alexandria Engineering Journal,* vol. 70, pp. 61–92, 2023/05/01/ 2023.
8. J. J. Justo, F. Mwasilu, J. Lee, and J.-W. Jung, "AC-microgrids versus DC-microgrids with distributed energy resources: A review," *Renewable and Sustainable Energy Reviews,* vol. 24, pp. 387–405, 2013/08/01/ 2013.
9. J. Kumar, A. Agarwal, and V. Agarwal, "A review on overall control of DC microgrids," *Journal of Energy Storage,* vol. 21, pp. 113–138, 2019/02/01/ 2019.
10. D. Jain and D. Saxena, "Comprehensive review on control schemes and stability investigation of hybrid AC-DC microgrid," *Electric Power Systems Research,* vol. 218, p. 109182, 2023/05/01/ 2023.
11. M. Pourbehzadi, T. Niknam, J. Aghaei, G. Mokryani, M. Shafie-khah, and J. P. S. Catalão, "Optimal operation of hybrid AC/DC microgrids under uncertainty of renewable energy resources:

A comprehensive review," *International Journal of Electrical Power & Energy Systems,* vol. 109, pp. 139–159, 2019/07/01/ 2019.

12. E. Planas, J. Andreu, J. I. Gárate, I. Martínez de Alegría, and E. Ibarra, "AC and DC technology in microgrids: A review," *Renewable and Sustainable Energy Reviews,* vol. 43, pp. 726–749, 2015/03/01/ 2015.
13. J. Saha and S. K. Panda, "Overview and comparative analysis of bidirectional cascaded modular isolated medium-voltage AC–low-voltage DC (MVAC-LVDC) power conversion for renewable energy rich microgrids," *Renewable and Sustainable Energy Reviews,* vol. 174, p. 113118, 2023/03/01/ 2023.
14. L. Ortiz, J. W. González, L. B. Gutierrez, and O. Llanes-Santiago, "A review on control and fault-tolerant control systems of AC/DC microgrids," *Heliyon,* vol. 6, no. 8, p. e04799, 2020/08/01/ 2020.
15. M. Monadi, M. Amin Zamani, J. Ignacio Candela, A. Luna, and P. Rodriguez, "Protection of AC and DC distribution systems Embedding distributed energy resources: A comparative review and analysis," *Renewable and Sustainable Energy Reviews,* vol. 51, pp. 1578–1593, 2015/11/01/ 2015.
16. M. Zolfaghari, G. B. Gharehpetian, M. Shafie-khah, and J. P. S. Catalão, "Comprehensive review on the strategies for controlling the interconnection of AC and DC microgrids," *International Journal of Electrical Power & Energy Systems,* vol. 136, p. 107742, 2022/03/01/ 2022.
17. W. Du, Q. Fu, X. Wang, and H. F. Wang, "Small-signal stability analysis of integrated VSC-based DC/AC power systems – A review," *International Journal of Electrical Power & Energy Systems,* vol. 103, pp. 545–552, 2018/12/01/ 2018.
18. A. A. Eajal, A. H. Yazdavar, E. F. El-Saadany, and K. Ponnambalam, "On the Loadability and Voltage Stability of Islanded AC–DC Hybrid Microgrids During Contingencies," *IEEE Systems Journal,* vol. 13, no. 4, pp. 4248-4259, 2019.
19. A. Ashok Kumar and N. Amutha Prabha, "A comprehensive review of DC microgrid in market segments and control technique," *Heliyon,* vol. 8, no. 11, p. e11694, 2022/11/01/ 2022.
20. X. Li *et al.*, "A Unified Control for the DC–AC Interlinking Converters in Hybrid AC/DC Microgrids," *IEEE Transactions on Smart Grid,* vol. 9, no. 6, pp. 6540-6553, 2018.
21. A. Agrawal, C. S. Nalamati, and R. Gupta, "Hybrid DC–AC Zonal Microgrid Enabled by Solid-State Transformer and Centralized ESD Integration," *IEEE Transactions on Industrial Electronics,* vol. 66, no. 11, pp. 9097-9107, 2019.
22. O. Husev, O. Matiushkin, D. Vinnikov, C. Roncero-Clemente, and S. Kouro, "Novel Concept of Solar Converter With Universal Applicability for DC and AC Microgrids," *IEEE Transactions on Industrial Electronics,* vol. 69, no. 5, pp. 4329-4341, 2022.
23. L. Che, M. Shahidehpour, A. Alabdulwahab, and Y. Al-Turki, "Hierarchical Coordination of a Community Microgrid With AC and DC Microgrids," *IEEE Transactions on Smart Grid,* vol. 6, no. 6, pp. 3042-3051, 2015.
24. Z. Fang *et al.*, "A Comprehensive Framework for Robust AC/DC Grid State Estimation Against Measurement and Control Input Errors," *IEEE Transactions on Power Systems,* vol. 37, no. 2, pp. 1067-1077, 2022.
25. A. Kirakosyan, E. F. El-Saadany, M. S. E. Moursi, A. H. Yazdavar, and A. Al-Durra, "Communication-Free Current Sharing Control Strategy for DC Microgrids and Its Application for AC/DC Hybrid Microgrids," *IEEE Transactions on Power Systems,* vol. 35, no. 1, pp. 140-151, 2020.
26. D. o. E. a. S. A. United Nations, *Multi Dimensional Issues In Internationational Electric Power Grid Interconnection.* New York: United Nations, 2006.

27. C. Wu, X. P. Zhang, and M. J. H. Sterling, "Economic Analysis of Power Grid Interconnections Among Europe, North-East Asia, and North America With 100% Renewable Energy Generation," *IEEE Open Access Journal of Power and Energy,* vol. 8, pp. 268-280, 2021.
28. Y. Wang, W. Hanr, and M. Hu, "The development trend of submarine cable projects for achieving optimal allocation of energy using international power grid interconnection," in *2017 EPTC Power Transmission and Transformation Technology Conference*, 2017, pp. 1–5.
29. K. Yu and P. van Son, "Review of trans-Mediterranean power grid interconnection: A regional roadmap towards energy sector decarbonization," *Global Energy Interconnection,* vol. 6, no. 1, pp. 115–126, 2023/02/01/ 2023.
30. C. Liang, J. Meng, C. Chen, and Y. Zhou, "A production-cost-simulation-based method for optimal planning of the grid interconnection between countries with rich hydro energy," *Global Energy Interconnection,* vol. 3, no. 1, pp. 23–29, 2020/02/01/ 2020.
31. T. G. Magallones and J. Govind Singh, "Impact of interconnections and renewable energy integration on the Philippine–Sabah Power Grid systems," *Global Energy Interconnection,* vol. 6, no. 3, pp. 253–272, 2023/06/01/ 2023.
32. T. J. Hammons *et al.*, "International High-Voltage Grids and Environmental Implications," *IEEE Power Engineering Review,* vol. 18, no. 8, pp. 4-4, 1998.
33. H. Kim, "Grid interconnections and decarbonization pathways for carbon neutrality of Northeast Asia," *Renewable Energy,* vol. 219, p. 119358, 2023/12/01/ 2023.
34. Y. Zhang, J. Xiao, Z. Zhou, Y. Zhao, H. Wan, and Y. Lin, "Research on the implementation of West Africa-North Africa grid interconnection using new electricity-water composite transmission technology," *Global Energy Interconnection,* vol. 4, no. 5, pp. 485–492, 2021/10/01/ 2021.
35. Y. Chi, G. Bai, and J.-H. Xu, "Security assessment of transnational power interconnection between China and neighboring countries," *Computers & Industrial Engineering,* vol. 171, p. 108336, 2022/09/01/ 2022.
36. T. J. Hammons *et al.*, "African electricity infrastructure interconnections and electricity exchanges," *IEEE Transactions on Energy Conversion,* vol. 15, no. 4, pp. 470-480, 2000.
37. H. G. Sarmiento and M. A. A. Rosales, "Issues associated with international power grid interconnections in Mexico," in *IEEE PES T&D 2010*, 2010, pp. 1-6.
38. T. Zhao *et al.*, "Index system and methods for comprehensive assessment of cross-border power grid interconnection projects," *Global Energy Interconnection,* vol. 3, no. 6, pp. 532–544, 2020/12/01/ 2020.
39. X. Lei, D. Wang, and W. Wang, "The scoping study of Kazakhstan-China-Republic of Korea power interconnection," *Global Energy Interconnection,* vol. 5, no. 5, pp. 484–500, 2022/10/01/ 2022.
40. C. Chen, C. Liang, Y. Zhou, C. Wang, J. Lu, and F. Xu, "A multi-level evaluation model for the recognition and assessment of impact factors associated with intercontinental power interconnection projects," *Global Energy Interconnection,* vol. 3, no. 1, pp. 30–42, 2020/02/01/ 2020.
41. S. Yamaguchi, T. Iitsuka, M. Osada, and R. Yokoyama, "Asian international grid connection and potentiality of DC superconducting power transmission," *Global Energy Interconnection,* vol. 1, no. 1, pp. 11–19, 2018/01/01/ 2018.
42. T. B. Scirocco *et al.*, "International benchmarking and new proposals on CBA methodology for the Italian transmission grid planning," in *2022 AEIT International Annual Conference (AEIT)*, 2022, pp. 1–5.
43. I. Australia, "BUSINESS CASE EVALUATION SUMMARY," AustraliaJan 2023 2023, Available: https://www.infrastructureaustralia.gov.au/projects/australia-asia-powerlink.

44. E. Cassetta, C. R. Nava, and M. G. Zoia, "EU electricity market integration and cross-country convergence in residential and industrial end-user prices," *Energy Policy,* vol. 165, p. 112934, 2022/06/01/ 2022.
45. M. P. Kirsten Westphal, Jacopo Maria Pepe, "Geopolitics of Electricity: Grids, Space and (political) Power," Stiftung Wissenschaft und Politik, Germany2022, Available: https://doi.org/10.18449/2022RP06/.
46. Z. Ji, X. Yu, W. Li, and D. Niu, "A multi-criteria decision-making framework for distributed generation projects investment considering the risk of electricity market trading," *Journal of Cleaner Production,* vol. 416, p. 137837, 2023/09/01/ 2023.
47. Y. J. T. Zidane, A. Johansen, and A. Ekambaram, "Project Evaluation Holistic Framework – Application on Megaproject Case," *Procedia Computer Science,* vol. 64, pp. 409–416, 2015/01/01/ 2015.
48. ElectraNet, " SA Energy Transformation RIT-T: Project Assessment Conclusions Report," South Australia13 February 2019 2019, Available: file:///C:/Users/WSA20150853/Downloads/SA-Energy-Transformation-PACR.pdf.
49. Y. Wu *et al.*, "An investment decision framework for photovoltaic power coupling hydrogen storage project based on a mixed evaluation method under intuitionistic fuzzy environment," *Journal of Energy Storage,* vol. 30, p. 101601, 2020/08/01/ 2020.
50. D. Darko, D. Zhu, M. Quayson, M. A. Hossin, O. Omoruyi, and A. K. Bediako, "A multicriteria decision framework for governance of PPP projects towards sustainable development," *Socio-Economic Planning Sciences,* vol. 87, p. 101580, 2023/06/01/ 2023.
51. M. Gimeno-Gutiérrez and R. Lacal-Arántegui, "Assessment of the European potential for pumped hydropower energy storage based on two existing reservoirs," *Renewable Energy,* vol. 75, pp. 856–868, 2015/03/01/ 2015.
52. K. J. Benes and C. Augustin, "Beyond LCOE: A simplified framework for assessing the full cost of electricity," *The Electricity Journal,* vol. 29, no. 8, pp. 48–54, 2016/10/01/ 2016.
53. M. M. Vanegas-Cantarero, S. Pennock, T. Bloise-Thomaz, H. Jeffrey, and M. J. Dickson, "Beyond LCOE: a multi-criteria evaluation framework for offshore renewable energy projects," *Renewable and Sustainable Energy Reviews,* vol. 161, p. 112307, 2022/06/01/ 2022.
54. B. Sowińska-Świerkosz and J. García, "A new evaluation framework for nature-based solutions (NBS) projects based on the application of performance questions and indicators approach," *Science of The Total Environment,* vol. 787, p. 147615, 2021/09/15/ 2021.
55. C. Lafleur, W. A. L. Truelove, J. Cousineau, C. E. Hiles, B. Buckham, and C. Crawford, "A screening method to quantify the economic viability of off-grid in-stream tidal energy deployment," *Renewable Energy,* vol. 159, pp. 610–622, 2020/10/01/ 2020.
56. O. Parkes, P. Lettieri, and I. D. L. Bogle, "Defining a quantitative framework for evaluation and optimisation of the environmental impacts of mega-event projects," *Journal of Environmental Management,* vol. 167, pp. 236–245, 2016/02/01/ 2016.
57. C. Ogwumike, H. Dawood, T. Ahmed, B. Gudlaugsson, and N. Dawood, "KPI evaluation framework and tools performance: a case study from the inteGRIDy Project," *Environmental Sciences Proceedings*, vol. 11, no. 1. https://doi.org/10.3390/environsciproc2021011023
58. T. Kealy, "A closed-loop renewable energy evaluation framework," *Journal of Cleaner Production,* vol. 251, p. 119663, 2020/04/01/ 2020.
59. Z. Xue-Wen, P. a. details, and F. N.-N. X. Z.; Ai-Hua Pei; Peng, Wei., "Comprehensive framework of major power project management based on system thinking," *Advances in Civil Engineering,* vol. 2022, 2022.
60. T. G. Ahmed, B. Gudlaugsson, C. Ogwumike, H. Dawood, M. Short, and N. Dawood, "Evaluation framework for Techno-economic analysis of energy system retrofit technologies," *Energy and Buildings,* vol. 286, p. 112967, 2023/05/01/ 2023.

61. M. Albawab, C. Ghenai, M. Bettayeb, and I. Janajreh, "Sustainability performance index for ranking energy storage technologies using multi-criteria decision-making model and hybrid computational method," *Journal of Energy Storage,* vol. 32, p. 101820, 2020/12/01/ 2020.
62. D. Olave-Rojas and E. Álvarez-Miranda, "Towards a complex investment evaluation framework for renewable energy systems: A 2-level heuristic approach," *Energy,* vol. 228, p. 120530, 2021/08/01/ 2021.

8 Space Energy Transfer

Solar PV panels cannot be used at night, but using space energy transfer, 100% use of the capacity of installed PV panels can be possible all day long. Space power transfer is an area that requires extensive exploration to realize this opportunity. Space power transfer is a type of long-range wireless power transfer. Wireless power transfer and energy harvesting is an area that is gaining further performance. This is being done from extremely short ranges of millimetres to the extremely long range over 100s of kilometers from space. Examples of this include everyday electronic devices charging through inductive coupling, medical applications such as in body battery charging, much longer-range power transfer over a few kilometers and then to the extreme long range from space. Wireless transmission has historically been employed for data transfer, but wireless power transmission has been confined to specific uses like mobile phone chargers. A critical concern in this context is power transmission efficiency, measured as the ratio of received power to transmitted power. For instance, delivering wireless power from a cell tower to a cell phone over a long distance results in poor efficiency, translating to significant financial losses. This efficiency metric, crucial in various Electrical Engineering applications, becomes particularly sensitive in wireless power transmission. Safety is another practical concern; if wireless power transmission poses biological, health or environmental hazards, public acceptance is at risk. Regulations have been instituted to protect against excessive exposure to wireless technologies. Electromagnetic compatibility is the third concern, exemplified by the National Radio Quiet Zone in the United States, which limits cell phone service to safeguard radio astronomical measurements from potential interference. Compliance with government regulations, including those from the Federal Communications Commission, is imperative for the development of wireless power transmission technology. Additionally, overcoming the substantial expenses related to the launch and

Md. R. Islam et al., *Universal Access to Sustainable, Affordable, and Reliable Electricity*, Synthesis Lectures on Sustainable Development,
https://doi.org/10.1007/978-3-032-19517-3_8

maintenance of space-based infrastructure poses extra challenges that must be resolved for the extensive adoption of space energy transfer systems [1–5].

Wireless Power Transmission Technologies

The technologies for wirelessly transmitting electrical power can be broadly categorized into three types: (a) low-frequency magnetic or electric fields, typically below 100 MHz, (b) high-frequency electromagnetic waves in bands above 100 MHz, including microwave frequencies, and (c) optical waves like infrared and visible lasers. Inductive coupling, utilizing low-frequency magnetic fields, and capacitive coupling, based on low-frequency electric fields, enable wireless power transmission for wireless mobile phone or electric vehicle charging. While inductive coupling efficiency drops with distance, laser propagation, demonstrated by German researchers in 2004 and the Naval Research Laboratory in 2019, allows for directional power transmission over long distances with better efficiency. However, optical waves face challenges like poor penetration capability and low conversion efficiency. The frequency range between low and optical frequencies emerges as a promising candidate for efficient wireless power transmission for different applications, with considerable research focused on microwave, millimeter wave, and Terahertz wave frequencies [4, 6].

Microwave power transmission technology has a significant history, with notable experiments spanning several decades [7–9]. In the 1960s, Brown demonstrated the groundbreaking transmission of microwave power from a ground station to a helicopter. At the same time, Peter Glaser has also explored the possibility of deploying large solar arrays in geostationary orbits to capture solar energy continuously and then convert it into microwave or laser beams for transmission to Earth. This approach has since garnered attention due to its potential to provide a consistent and clean energy source. Several proof-of-concept experiments, such as NASA's Wireless Power Transmission project, have demonstrated the feasibility of wireless energy transfer over short distances [10]. Raytheon's 1975 experiment achieved a 54% power transmission efficiency, and the Goldstone demonstration in the same year garnered attention. Programs like the Stationary High Altitude Relay Program (SHARP) in Canada and MIcrowave Lifted Airplane eXperiment (MILAX) in Japan aimed to provide microwave power to small aircraft in the 1980s and 1990s, respectively. In 1993, Japan conducted International Space Year-Microwave Energy Transmission (ISY-METS) experiments for microwave power transmission between spacecrafts [11]. A case study from 1997 to 2004 focused on wireless electricity transmission to the isolated village of Grand-Bassin in France. In 2009, a study explored using a car-borne power broadcaster to energize sensors on a bridge. Throughout the 2010s, numerous experiments, both ground-based and involving drones, showcased microwave power transmission. Presently, several companies are actively working on the development of commercial products for microwave power transmission.

Space Solar Energy Transfer Technologies

The concept of space solar power satellites (SSPS), introduced in 1968 [12], envisions harvesting solar power in Earth's geostationary orbit and wirelessly delivering it to the planet. The overall concept of space power transfer using SSPS is illustrated in Fig. 8.1.

This technology, if successful, could address the global energy crisis by providing steady GW level power. Unlike Earth-based solar harvesting, SSPS operates continuously, avoiding interruptions due to nightfall or weather conditions. The feasibility of SSPS remains under assessment, and wireless power transmission is a crucial aspect of this concept. Research efforts over the past decades, summarized in [13], focus on efficiently transmitting microwave power over approximately 36,000 km. The 1 to 10 GHz microwave frequency range, particularly around 2.45 and 5.8 GHz, is considered suitable for SSPS applications. Achieving efficient power transmission requires a large physical aperture (approximately 1 km^2) on the geostationary satellite's antenna panel. To minimize power loss and hazards, precise beam control is essential for accurately directing the wireless power beam to the designated ground station, considering the constant changes in the antenna panel's position and orientation.

SSPSs offer several significant advantages. Firstly, unlike terrestrial renewable energy sources, SSPS ensures uninterrupted power generation 24 h a day, irrespective of weather conditions. The solar irradiance in space, being 40% stronger than on the ground, enhances the efficiency of power generation. Additionally, SSPS allows for directed power delivery to different locations on demand, eliminating the need for extensive power lines. From an environmental standpoint, SSPS emits no greenhouse gases and produces no hazardous waste during power generation. Furthermore, the system is less vulnerable to terrestrial natural disasters, such as earthquakes, providing a robust and continuous power supply [14].

Advantages and Challenges of Space Power Transfer

However, the deployment of SSPS faces significant challenges. The development of space transportation technology for launching bulky materials, assembling large structures in orbit, and ensuring in-orbit operations and maintenance over extended periods poses a technological hurdle. Efficient and safe generation, transmission, and reception of power also demand substantial technological advancements. Safety concerns arise from potential adverse effects of microwaves/lasers on various aspects, including human health, the

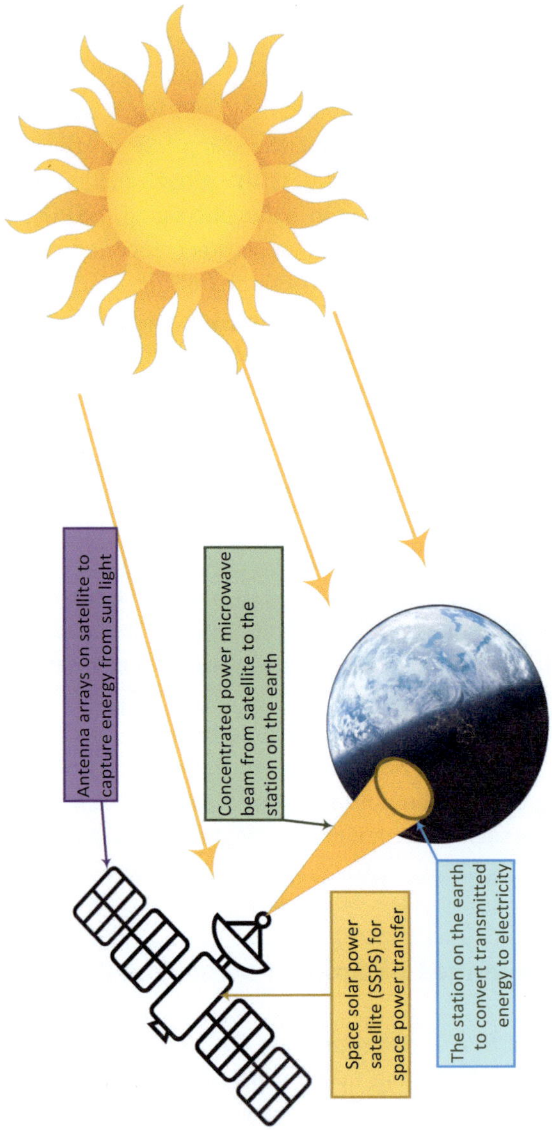

Fig. 8.1 The overall concept of space power transfer using SSPS

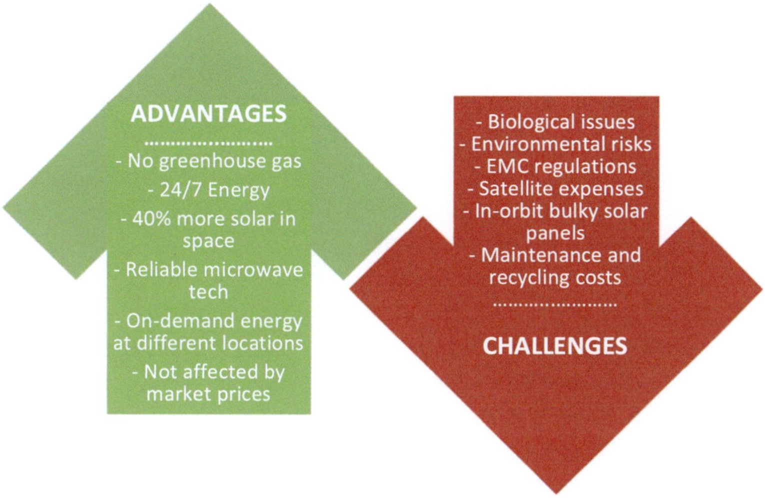

Fig. 8.2 The advantages and challenges of space power transfer

atmosphere, ionosphere, airplanes, and electronic appliances. Measures must be implemented to prevent, withstand, and repair damage caused by space debris, solar flares, and other environmental factors. Additionally, economic considerations play a role, with the need to reduce costs associated with space transportation, orbital assembly, operation and maintenance, and removal to make SSPS competitive with other energy sources [15]. Securing microwave frequencies, orbital slots, and terrestrial rectenna sites further adds complexity to the economic challenges of SSPS implementation. Comprehensive research should be conducted to overcome the challenges of space power transfer to facilitate the use of this green energy resource. The advantages and challenges of space power transfer are summarized in Fig. 8.2.

For the cost and emission comparison of SSPS, two designs are evaluated by NASA [16]: the innovative heliostat swarm (RD1) and the mature planar array (RD2), both intended to deliver 2 GW of power to the grid by 2050. With the current technological trend, the cost of electricity for RD1 is \$0.61/kWh and for RD2 is \$1.59/kWh, with launch costs being the most significant factor, accounting for 71% of RD1's and 77% of RD2's total costs. However, suppose launch costs decrease to \$50m per launch, solar efficiency improves to 50%, hardware lifetime increases to 15 years, and other productivities are achieved. In that case, the costs could drop significantly to \$0.04/kWh for RD1 and \$0.08/kWh for RD2. Greenhouse gas emissions could also be reduced by 86–89%, making these systems more environmentally friendly [16].

Space Power Transfer Case Studies

The accomplishment of wireless power transfer was realized through the Microwave Array for Power-transfer Low-orbit Experiment (MAPLE), a collection of flexible and lightweight microwave power transmitters aboard the Space Solar Power Demonstrator (SSPD-1) [17]. Launched as part of the California Institute of Technology's Space Solar Power Project (SSPP) in January 2023, SSPD-1 aims to harvest solar power in space and transmit it to Earth. MAPLE's successful power transmission to space receivers has been confirmed, including its programmable ability to direct energy toward Earth. The experiment, conducted in the harsh space environment, showcased MAPLE's resilience to temperature swings and solar radiation. This prototype's capabilities, including rapid energy direction adjustment on a nanosecond scale without moving parts, pave the way for larger-scale SSPP units, demonstrating the potential for precise and controlled wireless energy transmission in space and to Earth-based receivers. It uses an array of flexible, lightweight microwave power transmitters to beam energy to desired locations. For SSPP to become feasible, energy transmission arrays must be lightweight, flexible, and cost-effective [3]. MAPLE's successful demonstration of wireless power transfer in 2023 marks a significant milestone in the progress of this technology. The experiment faces the challenges of stability in the space environment, including temperature swings and solar radiation, but serves as a promising case study for advancing space power transfer capabilities.

SOLARIS is another technology for transferring energy from space. SOLARIS aims to lay the groundwork for a potential space solar power development program in 2025 by establishing technical, political, and programmatic viability. Through limited initial investment, SOLARIS will conduct studies and technology developments in collaboration with European industry to mature technical feasibility, assess benefits, implementation options, commercial opportunities, and risks of space power transfer. It addresses environmental, health, safety issues, and regulation challenges while advancing key technologies like high-efficiency solar cells, wireless power transmission, and robotic in-orbit assembly. The visionary SSPS design in this project, inspired by a butterfly, incorporates V-shaped solar panels with integrated antennas, providing a continuous 23 MW of power for lunar surface operations. The lunar station, located at an Earth-Moon Lagrange point, serves as a gateway between Earth and Moon operations, potentially becoming a tourist destination and prototype for future space settlements [18]. Concluding in August 2022, studies on Solaris's cost and benefits suggest its potential to provide competitively priced electricity in Europe by 2040, displacing fossil fuels and enhancing energy security. When deployed at scale, Solaris offers substantial environmental, economic, and strategic advantages. Despite requiring further technological developments, including efficient gigawatt-scale power collection and safe delivery to Earth, Europe possesses foundational building blocks. To leverage this game-changing capability in addressing climate concerns, awareness raising and increased investments in technology R&D are crucial. While

SOLARIS appears promising as a complementary energy source, both technical and non-technical challenges must be addressed before a final decision in 2025. A new program, SOLARIS, proposed by the European Space Agency (ESA) between 2023 and 2025, aims to further investigate SBSP's technical feasibility and costs, providing essential insights for an informed decision in 2025 on pursuing this transformative capability. The LCOE for different numbers of stations is provided in Fig. 8.3 [15]. As seen, space-based power stations and launch are the main expenses of SSPS, contributing 82%, 89%, and 90% to the total LCOE for 3, 4, and 5 power stations cases for SSPS, respectively.

The China Academy of Space Technology (CAST) is also set to conduct a groundbreaking "Space High Voltage Transfer and Wireless Power Transmission Experiment" in 2028 [19]. The satellite planned for this experiment is designed to generate 10 kWs and will carry essential components, including a solar cell array, microwave transmitting antenna, low-power laser transmission payload, and a transmitting array. The test aims to demonstrate power transmission capabilities over distances of 400 km from orbit. Following the 2028 test, phase 2 is scheduled for 2030, involving a launch into geostationary

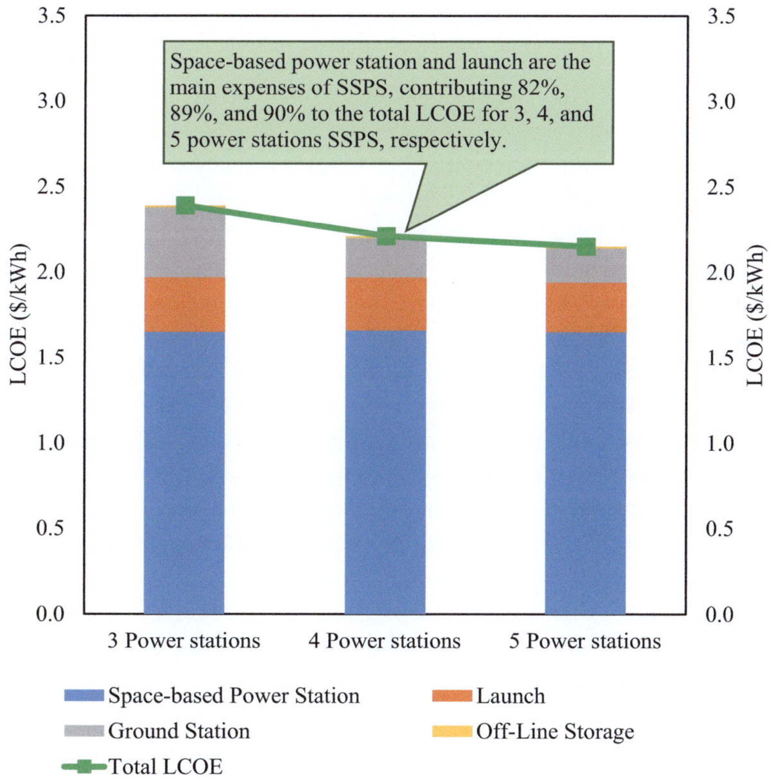

Fig. 8.3 The LCOE for different numbers of stations

orbit and requiring accurate energy transmission over a substantial distance of 35,800 km to Earth, generating up to one megawatt. This mission includes larger transmission arrays, medium power laser transmission, and on-orbit assembly. Phases 3 and 4, planned for 2035 and 2050, present formidable challenges with significant increases in energy generation (10 MW and 2 GW), demanding advanced orbital assembly capabilities, precise beam steering accuracy, and expansive transmission architectures. These later phases envision spacecraft with transmission arrays exceeding 100 m and around 1 km, marking ambitious milestones in space-based wireless power transmission technology.

In addition, Russian space systems, a subsidiary of Roscosmos, has unveiled an advanced space power plant project designed to harness solar energy and transmit it to Earth [20]. The system comprises a 70-square-meter unmanned spacecraft in space, collecting and storing solar energy, and a ground station with an antenna array to wirelessly receive and distribute the accumulated energy. The energy transfer utilizes a laser beam with minimal energy loss, enhancing efficiency. Solar panels in space benefit from increased effectiveness, avoiding atmospheric energy loss and cloud interference. The spacecraft is equipped with energy storage capabilities, a control unit for balanced distribution, and software for precise targeting of receiving antennas. While the company hasn't specified energy loss during the transition or mentioned a working prototype, this initiative aligns with experiments on laser-based energy transfers, a priority research area at the Russian section of the International Space Station, suggesting potential synergy between these endeavors.

The ongoing debate over achieving 100% terrestrial renewables by 2050 faces challenges linked to rising global population and energy demands including for the moon and mars mission [8, 21, 22]. Balancing energy equality and preventing widespread poverty while avoiding escalating carbon emissions requires innovative solutions. Although nuclear fission and future terrestrial fusion hold promise, they encounter public opposition and operational delays. Space solar power emerges as a potential solution, tapping into the abundant energy of the Sun. However, space power transfer development has been hindered by scalability issues and high start-up costs [23]. The CASSIOPeiA project aims to overcome these challenges by introducing a non-scaling paradigm. CASSIOPeiA eliminates rotating interfaces, allowing subsystems to share a modular structure, enabling scalable power density from sub-megawatt to gigawatt systems. With unique beam-steering capabilities, CASSIOPeiA facilitates baseload delivery from alternative orbits, offering a lower-risk and lower-cost entry point into the era of space power transfer through near-term stratospheric station-keeping platforms [24]. This approach emphasizes adaptability, retrieval, servicing, and knowledge transfer, potentially revolutionizing the feasibility of space power transfer deployment.

A detailed and deep research should also be conducted on the design of resilient PV panel [25] and the adaptive transfer frequency tuning and managing hazards to the environment due to the transfer of high amounts of energy from space to the Earth. This is because the amount of transferred energy and the different frequencies are different [26].

Transmission Antenna Technologies

Space-based energy transfer often relies on highly efficient and specialised antennas designed to transmit power across long distances. One prominent type of antenna used for this purpose is the Rectenna (Rectifying Antenna), which functions by converting received microwave or radio frequency signals into electrical energy through a combination of rectification and electromagnetic resonances [27]. These antennas are equipped with arrays of rectifying diodes that transform the alternating current induced by the incoming electromagnetic waves into direct current, suitable for powering devices or charging batteries. The advantage of rectennas lies in their ability to capture and convert energy from microwave beams sent from space-based transmitters, enabling the wireless transmission of power to remote locations, such as space stations, without the need for physical cables. Space-based energy transfer to ground stations typically employs two main types of antennas: the space-based transmitting antenna and the ground-based receiving antenna.

- Space-based transmitting antenna

In space energy transfer systems, high-power microwave or laser beams are often used to transmit energy from space-based platforms, such as satellites or solar power stations, to the Earth's surface [28]. For microwave-based systems, phased-array antennas are commonly used. Phased-array antennas allow precise control of the direction of the transmitted beam by adjusting the phase of individual elements within the array. This control is crucial to focus the energy beam accurately toward the ground-based receiving station. These antennas can be planar (i.e., patch and slot antennas), making them suitable for deployment on satellites or other space platforms [29, 30].

- Ground-based receiving antenna (rectenna)

On the ground, a specialized receiving antenna known as a rectenna is used to capture the transmitted energy and convert it into usable electricity [28]. A rectenna consists of an array of dipole antennas combined with rectifying diodes. The dipole antennas capture the incoming microwave energy, and the rectifying diodes convert the alternating current generated by the antennas into direct current. The rectenna array is typically large to cover a significant area and capture as much of the transmitted energy as possible. Due to the size requirements, ground-based rectennas are often installed in remote or open areas.

These two types of above-mentioned antennas work in conjunctions to enable the transfer of energy from space to the ground. The space-based transmitting antenna generates a focused beam of energy that is directed toward the ground-based rectenna. The rectenna then captures and converts this energy into usable electricity for various applications, ranging from powering remote areas to supporting space missions and beyond.

Future Trends

Looking ahead, several trends are likely to shape the future of space energy transfer. Advances in materials science and manufacturing techniques could lead to the development of more efficient and lightweight solar arrays, reducing the cost of launching and deploying space-based systems. Research into beamforming and beam steering technologies could enhance the precision of energy transmission, mitigate safety concerns and optimizing energy delivery by pointing the beam to the desired direction only. Moreover, collaborative international efforts and public–private partnerships may play a crucial role in driving research and development in this field. As the global demand for clean and sustainable energy sources continues to grow, space energy transfer holds the promise of providing a consistent and virtually unlimited power source, making it an important avenue for exploration in the coming years.

References

1. X. Wang and M. Lu, "Wireless power transmission based on retro-reflective beamforming technique," *Space Solar Power and Wireless Transmission,* 2023/10/16/ 2023.
2. J. C. Mankins, "New directions for space solar power," *Acta Astronautica,* vol. 65, no. 1, pp. 146–156, 2009/07/01/ 2009.
3. E. C. Warmann *et al.*, "An ultralight concentrator photovoltaic system for space solar power harvesting," *Acta Astronautica,* vol. 170, pp. 443–451, 2020/05/01/ 2020.
4. S. Sasaki, K. Tanaka, and K. i. Maki, "Microwave Power Transmission Technologies for Solar Power Satellites," *Proceedings of the IEEE,* vol. 101, no. 6, pp. 1438–1447, 2013.
5. P. Malaviya, V. Sarvaiya, A. Shah, D. Thakkar, and M. Shah, "A comprehensive review on space solar power satellite: an idiosyncratic approach," *Environmental Science and Pollution Research,* vol. 29, no. 28, pp. 42476–42492, 2022/06/01 2022.
6. P. Jaffe and J. McSpadden, "Energy Conversion and Transmission Modules for Space Solar Power," *Proceedings of the IEEE,* vol. 101, no. 6, pp. 1424-1437, 2013.
7. Y. Yang, Y. Zhang, B. Duan, D. Wang, and X. Li, "A novel design project for space solar power station (SSPS-OMEGA)," *Acta Astronautica,* vol. 121, pp. 51–58, 2016/04/01/ 2016.
8. B. K. Durant, I. R. Sellers, and B. Rout, "Chapter Seven - Perovskite solar cells on the horizon for space power systems," in *Photovoltaics for Space*, S. G. Bailey, A. F. Hepp, D. C. Ferguson, R. P. Raffaelle, and S. M. Durbin, Eds.: Elsevier, 2023, pp. 175–195.
9. S. Gosavi *et al.*, "A Review on Space Based Solar Power," *Journal of Thermal Energy Systems,* vol. 6, pp. 16–24, 02/11 2021.
10. A. K. Porter *et al.*, "Demonstration of Electromagnetic Formation Flight and Wireless Power Transfer," *Journal of Spacecraft and Rockets,* vol. 51, no. 6, pp. 1914-1923, 2014.
11. M. Mori, H. Kagawa, and Y. Saito, "Summary of studies on space solar power systems of Japan Aerospace Exploration Agency (JAXA)," *Acta Astronautica,* vol. 59, no. 1, pp. 132–138, 2006/07/01/ 2006.
12. P. E. GLASER, "Power from the Sun: Its Future," *SCIENCE,* vol. 162, no. 3856, pp. 857–861, 22 Nov 1968 1968.

13. B. Strassner and K. Chang, "Microwave Power Transmission: Historical Milestones and System Components," *Proceedings of the IEEE,* vol. 101, no. 6, pp. 1379-1396, 2013.
14. JAXA. *About the SSPS*. Available: https://www.kenkai.jaxa.jp/eng/research/ssps/ssps-ssps.html.
15. M. A. Marshall, R. G. Madonna, and S. Pellegrino, "Investigation of Equatorial Medium Earth Orbits for Space Solar Power," *IEEE Transactions on Aerospace and Electronic Systems,* vol. 58, no. 3, pp. 1574-1592, 2022.
16. E. G. Erica Rodgers, Jordan Sotudeh, Carie Mullins, Amanda Hernandez, Hanh Nguyen Le, Phil Smith, and Nikolai Joseph, "Space-based solar power," NASA, USA2024, Available: https://www.nasa.gov/wp-content/uploads/2024/01/otps-sbsp-report-final-tagged-approved-1-8-24-tagged-v2.pdf.
17. CalTech. (2023, 1 June 2023). *In a First, Caltech's Space Solar Power Demonstrator Wirelessly Transmits Power in Space.* Available: https://www.caltech.edu/about/news/in-a-first-caltechs-space-solar-power-demonstrator-wirelessly-transmits-power-in-space.
18. T. E. S. Agency. (2023). *Solaris.* Available: https://www.esa.int/Enabling_Support/Space_Engineering_Technology/SOLARIS.
19. D. S. H. X. W. Xin, "Retro-directive microwave power beam steering technology for space solar power station," *Chinese Space Science and Technology,* vol. 2022, no. 05, pp. 91-102.
20. (2022). *Russian Space Company Develops Method for Effective Transfer of Solar Energy to Any Spot on Earth.* Available: https://sputnikglobe.com/20220105/russian-space-company-develops-method-for-effective-transfer-of-solar-energy-to-any-spot-on-earth-1092044555.html.
21. A. Vaughan, "Solar panels in space could help power the UK by 2039," *New Scientist,* vol. 251, no. 3349, p. 17, 2021/08/28/ 2021.
22. A. Baraskar, Y. Yoshimura, S. Nagasaki, and T. Hanada, "Space solar power satellite for the Moon and Mars mission," *Journal of Space Safety Engineering,* vol. 9, no. 1, pp. 96–105, 2022/03/01/ 2022.
23. D. L. Dotson, J. Eddy, and P. Swan, "Climate action and growing electricity demand: Meeting both challenges in the 21st century with space-based solar power delivered by space elevator," *Acta Astronautica,* vol. 198, pp. 761–766, 2022/09/01/ 2022.
24. I. Cash, "CASSIOPeiA – A new paradigm for space solar power," *Acta Astronautica,* vol. 159, pp. 170–178, 2019/06/01/ 2019.
25. G. Fan, B. Duan, Y. Zhang, X. Li, and X. Ji, "Full-spectrum selective thin film based photonic cooler for solar cells of space solar power station," *Acta Astronautica,* vol. 180, pp. 196–204, 2021/03/01/ 2021.
26. J. Straub, "Does the use of space solar power for in-space activities really make sense: An updated economic assessment," *Space Policy,* vol. 31, pp. 21–26, 2015/02/01/ 2015.
27. E. Donchev *et al.*, "The rectenna device: From theory to practice (a review)," *MRS Energy & Sustainability,* vol. 1, p. E1, 2014.
28. Y. Murao and T. Takano, "An investigation on the design of a transmission antenna and a rectenna with arrayed apertures for microwave power transmission," *Electronics and Communications in Japan (Part I: Communications),* vol. 83, no. 2, pp. 1–9, 2000.
29. S. Abulgasem, F. Tubbal, R. Raad, P. I. Theoharis, S. Lu, and S. Iranmanesh, "Antenna designs for CubeSats: A review," *IEEE Access,* vol. 9, pp. 45289-45324, 2021.
30. F. E. Tubbal, R. Raad, and K.-W. Chin, "A survey and study of planar antennas for pico-satellites," *IEEE Access,* vol. 3, pp. 2590-2612, 2015.

9 Discussion and Conclusion

Research Gaps Mapping

Figure 9.1 shows the mapping of all research gaps that are identified and discussed in this book on different categories associated with power systems including generation, delivery, operation, and consumption. This mapping help decision-makers and researchers to understand the main impact of each research gap in each section of the industry. As discussed in this book, the proposed topics align with the vision of unlocking universal access to sustainable, affordable, and reliable (Threeable) electricity.

Some research gaps are associated with only one section of electricity industry, while some other research gaps are related to more than one section of electricity industry. For example, you can see nuclear power on the generation side or electric vehicles on the consumption side. Whereas energy storage are associated to all categories, as the outcome of the research in these topics will impact the way of management, investment, and operation in those sections. The topics of zero net emission planning, electricity price, and real cost of renewables are also mainly pertaining to both generation and delivery and operation of electricity industry. Also, a red line at the boundaries of all categories represents cyber security necessary for all areas in the new era of power systems.

Research Topics Summary

Figure 9.2 shows the critical research gaps and key questions for enabling energy systems to move towards sustainable, affordable, and reliable energy systems. In the topic of Reliability and Resiliency, several research gaps emerge. Firstly, there's a need to explore "reliability as a service" models and understand the marginal cost of reliability in electricity markets, particularly concerning renewable and energy storage systems.

Md. R. Islam et al., *Universal Access to Sustainable, Affordable, and Reliable Electricity*, Synthesis Lectures on Sustainable Development,
https://doi.org/10.1007/978-3-032-19517-3_9

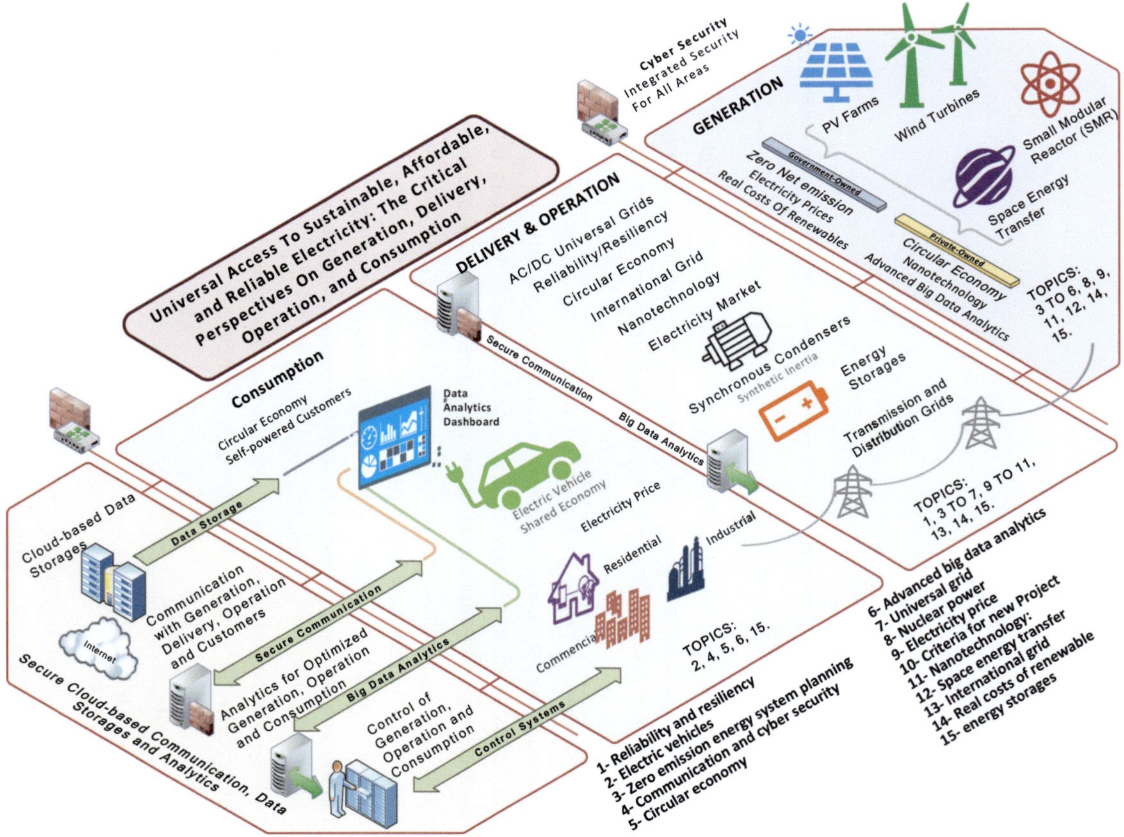

Fig. 9.1 The mapping of the proposed research gaps in generation, delivery, operation, and consumption for the sustainable, affordable, and reliable (Threeable) electricity

Secondly, addressing the uncertainties and faults inherent in renewables and integrating them into reliability assessments and restoration strategies is crucial for enhancing grid resilience. Additionally, the development of fast reliability analysis methods is imperative for efficient power system planning, allowing for timely decision-making and adaptation to dynamic conditions.

Turning to EVs, significant research gaps exist. Firstly, the impact of ultra-fast charging on power grids and its implications for grid stability require thorough investigation, considering the increasing adoption of EVs. Moreover, optimizing incentives to shape EVs' loading consumption is essential for managing grid stress and ensuring efficient energy utilization. Lastly, exploring the development of autonomous EV shared economies holds promise for reducing risks associated with individual ownership while effectively managing resources and infrastructure.

In the domain of energy storage, several critical research gaps demand attention. Firstly, there's a pressing need to investigate effective integration and control strategies for energy storage systems to foster a cleaner, more resilient, and economically efficient electricity market. Additionally, leveraging energy storage control mechanisms to alleviate grid congestion, enhance reliability, and manage peak demand represents a significant research frontier. Lastly, determining the optimal portfolio of energy storage technologies to achieve zero emissions targets requires sophisticated modeling and analysis, considering various factors such as cost-effectiveness, scalability, and environmental impact. Addressing these research gaps is essential for advancing the deployment and integration of energy storage solutions into the evolving energy landscape.

In the area of Zero Emissions Energy System Planning, significant research gaps exist, particularly concerning the demand for materials like rare earth elements (REEs). A comprehensive solution is needed to mitigate this demand while ensuring sustainable energy systems. Additionally, there is a need to explore the integration of new technologies such as superconducting cables, synchronous condensers, and power electronic transformers, as well as the implementation of peer-to-peer transactions, to optimize energy distribution and utilization.

In the domain of Circular Economy, several research gaps warrant attention. Firstly, there is a lack of clarity regarding the percentage of recycled materials that manufacturers should incorporate into their products, and understanding the projected trend by 2050 is crucial for sustainable resource management. Furthermore, investigating the value chain of circular economy practices for energy storage systems and renewable energy sources is essential for maximizing their efficiency and minimizing environmental impact. Additionally, extending the lifetime of grid components and exploring opportunities for repurposing them could significantly contribute to resource conservation and circularity.

Regarding Nuclear Power, there are notable gaps in research that demand exploration. Firstly, accurate evaluation methods are needed to assess the lifetime emissions and levelized cost of energy (LCOE) of small modular reactors (SMRs) compared to other

Fig. 9.2 The critical research gaps and key questions for universal access to sustainable, affordable, and reliable electricity

renewables, taking into account factors like grid expansion costs, stability, and the integration of energy storage solutions. Additionally, addressing the psychological aspects of communities residing near nuclear power plants is essential for understanding public perceptions, building trust, and ensuring effective communication about safety measures and risk mitigation strategies. These research gaps are critical for informing policy decisions and technological advancements in the nuclear energy sector.

In the realm of electricity prices, several research gaps exist, particularly regarding the impact of evolving electricity markets, advancements in power system management, and the implementation of new standards for renewables and energy storage technologies. Additionally, there is a need to explore strategies for redesigning power grids in Australia, aiming to mitigate their disproportionate costs in electricity bills, which are often influenced by practices more suitable for densely populated regions. This entails investigating innovative approaches to grid design and management that align with Australia's unique geographic and demographic characteristics.

Concerning the Universal Power Grid, significant research gaps revolve around understanding the potential benefits and challenges associated with its implementation, along with the economic, regulatory, and policy considerations involved. Furthermore, there is a need to delve into converter technologies, communication-less control strategies, and optimal grid architecture to facilitate seamless integration and operation. Additionally, exploring the implications of universal power grids on stability, protection, and fault-tolerant control systems is crucial for ensuring reliability and resilience in the face of evolving energy landscapes.

In the domain of Space Energy transfer, several challenges must be addressed to realize its potential. Research gaps include overcoming biological and environmental risks associated with space-based energy systems, managing electromagnetic compatibility issues, and addressing concerns related to satellite maintenance and recycling costs. Furthermore, there is a need for innovative designs of resilient and compact in-orbit photovoltaic panels, along with adaptive frequency tuning for efficient energy transfer. Additionally, advancements in precise beamforming and beam steering technologies are essential to optimize energy transmission from space to terrestrial receivers, further underscoring the complexity and interdisciplinary nature of this field.

The research gaps identified in the domain of International Grid pertain to the absence of a comprehensive evaluation framework that holistically assesses the risks and benefits associated with international grids across various dimensions. Firstly, there's a notable gap in the development of an evaluation framework that considers economic, environmental, social, cultural, legal, and political aspects. Additionally, critical matters such as reliability, international energy market volatility, and interconnection costs remain inadequately assessed. Conflict arising from inappropriate allocation of costs and benefits of interconnection further exacerbates the complexity of international grid projects, highlighting the need for robust assessment methodologies to address these challenges.

In the topic of assessing new projects in energy systems, significant research gaps exist regarding the establishment of a comprehensive assessment framework. Particularly, there's a need for the development of performance indicators that encompass a wide array of criteria essential for ensuring the effectiveness and viability of new projects in the long term. These criteria include technical feasibility, economic viability, political acceptability, legal compliance, human resource development, environmental impact, and data security. The absence of such a framework hampers the ability to accurately evaluate the potential success and sustainability of new energy projects, underscoring the urgency for research efforts aimed at addressing this gap.

On the side of Real Costs of Renewables is also critical to conduct research on developing a framework to evaluate the cost of renewables, including their intermittency, supporting technologies such as energy storage, inertia support, communication, and lifecycle environmental emission and pollution.

The Interconnections of Research Gaps with Each Other

The identified research gaps have interaction and synergy with each other as they are associated with the same electricity system. Researchers can identify new research ideas by considering the interaction between the research gaps identified and discussed in this book. For example, the reliability and resiliency of energy systems are fundamental prerequisites for the successful integration of renewable energy sources and supporting technologies such as energy storage and electric vehicles (EVs). Without robust reliability measures, uncertainties in renewable generation and potential grid instabilities could undermine the feasibility and scalability of renewable energy deployment. Therefore, research efforts aimed at enhancing reliability and resiliency directly impact on the feasibility and success of initiatives across various sectors.

Moreover, the reliability of energy systems is intricately linked to the efficiency and effectiveness of supporting technologies. For instance, energy storage systems play a crucial role in bolstering grid resiliency by providing backup power during outages and mitigating intermittency issues associated with renewables. Hence, research endeavors focused on enhancing reliability and resiliency directly influence the development and optimization of supporting technologies as well, thereby addressing critical gaps in their integration and operation.

Furthermore, the reliability and resiliency of energy systems have profound implications for other research domains, such as net-zero emissions energy system planning and international grid integration. A reliable and resilient energy infrastructure forms the backbone of sustainable energy systems, facilitating the effective deployment and management of zero emission technologies. Additionally, robust reliability measures are essential for ensuring the stability and viability of international grid projects, which rely on seamless cross-border energy exchange. Therefore, research initiatives aimed at improving reliability and resiliency contribute to addressing key challenges and gaps across diverse areas of energy systems planning and integration.

The research gap concerning EVs plays a pivotal role in influencing various aspects of the broader research landscape in energy systems. Firstly, the impact of ultra-fast charging on power grids and grid stability is a significant concern, which is related to the "Reliability and Resiliency" research gaps. Moreover, optimizing incentives to shape EVs' loading consumption is also connected to zero-emission energy system planning, as the cost and benefit of such arrangement are essential in the planning.

Furthermore, the research gap in EVs intersects with the domain of energy storage. Electric vehicle batteries represent a form of distributed energy storage that can potentially contribute to grid stability and resilience. Investigating the integration of EV batteries into energy storage systems and exploring control strategies to leverage their storage capacity during peak demand periods can enhance grid reliability and mitigate the need for additional infrastructure investments. Additionally, examining the optimal portfolio of energy storage technologies, including EV batteries, is essential for achieving zero emissions targets and ensuring cost-effectiveness in the transition to renewable energy sources.

Moreover, the research gap in EVs has implications for the universal power grid. As EV adoption continues to rise globally, the electrification of transportation introduces new dynamics into the energy system. Understanding the impact of EVs on grid infrastructure and assessing the feasibility of incorporating EVs into the universal power grid are critical steps towards achieving seamless integration and operation. Furthermore, exploring the potential synergies between EVs and renewable energy generation can pave the way for a more sustainable and resilient energy future.

The research gap of "Energy Storage" plays a pivotal role in influencing various aspects of the broader research landscape towards Threeable electricity systems. Energy storage systems can serve as a buffer to mitigate the intermittency of renewable energy sources, thereby contributing to a more stable and dependable grid infrastructure. Understanding effective integration and control strategies for energy storage systems is essential for fostering grid resilience against uncertainties and faults inherent in renewables, as mentioned in the research gap of "Reliability and Resiliency".

Moreover, energy storage technologies have significant implications for other research areas such as EVs and Zero Emissions Energy System Planning. In the context of EVs, the impact of ultra-fast charging on power grids necessitates a thorough investigation, highlighting the importance of energy storage systems in managing grid stress and ensuring efficient energy utilization. Additionally, in Zero Emissions Energy System Planning, determining the optimal portfolio of energy storage technologies is essential for achieving zero emissions targets and fostering a cleaner, more sustainable energy market.

Furthermore, the deployment of energy storage solutions also intersects with research gaps in Circular Economy and International Grid. Investigating the value chain of circular economy practices for energy storage systems is crucial for maximizing their efficiency and minimizing environmental impact. Energy storage technologies can facilitate the efficient utilization of renewable energy resources across international grids, emphasizing the interconnectedness of global energy systems and the importance of comprehensive evaluation frameworks, as highlighted in the research gap of the International Grid.

The research gap of "Zero Emissions Energy System Planning" has significant implications for the other research gaps identified in the text. Firstly, addressing the demand for materials like REEs is essential not only for ensuring sustainable energy systems but also for supporting the development and deployment of supporting technologies such as energy storage systems and EVs. Many of these technologies rely on rare materials for their production, and mitigating the demand for such materials is crucial for achieving widespread adoption and scalability, which is again related to the circular economy topic. Secondly, the integration of new technologies, such as superconducting cables and power electronic transformers, into energy systems requires careful planning and coordination. Research into their integration and implementation is closely tied to the overarching goal of achieving zero emissions energy systems and reliability and resiliency.

Furthermore, exploring the implementation of peer-to-peer transactions in energy systems can have ripple effects across various domains. By decentralizing energy distribution and enabling direct transactions between consumers and producers, peer-to-peer transactions can promote energy democratization, enhance grid flexibility, and incentivize the adoption of renewable energy technologies. This research gap intersects with others, such as EVs, energy storage, universal power grids, and international grids, as it pertains to the evolution of energy markets and regulatory frameworks on both local and global scales.

The research gap concerning the Circular Economy also interacts with the rest of the identified research gaps within the energy sector. Firstly, the Circular Economy research gap intersects with the Energy Storage and Zero Emissions Energy System Planning gaps. By investigating the value chain of circular economy practices for energy storage systems and renewable energy sources, researchers can identify opportunities to maximize efficiency and minimize resource consumption. This includes exploring strategies for recycling and repurposing materials used in energy storage systems, such as batteries, to extend their lifespan and reduce the demand for new materials.

Secondly, the Circular Economy gap is intertwined with the Real Costs of Renewables gap. Understanding the percentage of recycled materials that manufacturers should incorporate into renewable energy technologies is crucial for assessing their true costs. Incorporating recycled materials can not only reduce production costs but also mitigate the environmental impacts associated with raw material extraction and processing. Additionally, extending the lifetime of grid components through circular economy practices can reduce the need for frequent replacements, thus lowering overall lifecycle costs.

Furthermore, the Circular Economy gap impacts research areas such as Reliability and Resiliency and Electricity Prices. By exploring opportunities for repurposing and reusing grid components, researchers can contribute to enhancing the reliability and resilience of electricity grids to some extent. Additionally, incorporating recycled materials into renewable energy technologies can potentially lower production costs, leading to reduced electricity prices for consumers. The research gap concerning nuclear power holds significant implications for the broader understanding of the real costs of renewables. Nuclear power presents an alternative to traditional fossil fuels and renewable energy sources in terms of its potential to provide consistent and large-scale electricity generation. However, the evaluation of nuclear power's real cost extends beyond its direct generation expenses. Firstly, accurate assessment methods are required to compare the lifetime emissions and LCOE of SMRs with other renewables. Due to these reasons, the topic of nuclear power influences other topics such as reliability and resiliency, and energy storage. Furthermore, considering nuclear power as a new type of renewable energy sheds light on the broader spectrum of energy options available for achieving emissions reduction targets and ensuring energy security. This makes this topic related to Zero Emission Energy System Planning.

The research gap concerning "Electricity Prices" connects with other identified research gaps in various domains within the energy sector. In the domain of Reliability and Resiliency, electricity prices directly impact the cost–benefit analysis of reliability-enhancing measures. Similarly, in the context of EVs, electricity prices significantly influence the economic viability of electric vehicle adoption. Fluctuation in electricity prices can impact the cost of charging EVs, thereby affecting consumer demand and adoption rates. Research into electricity prices can inform strategies for optimizing EV charging patterns to minimize grid stress during peak demand periods, contributing to grid stability and reliability. Moreover, electricity prices are intricately linked to the economics of energy storage systems. Research into electricity prices can shed light on the cost-effectiveness of energy storage technologies, particularly in relation to arbitrage opportunities and peak-shaving strategies. Additionally, electricity prices influence the competitiveness of nuclear power and renewable energy technologies and their integration into the energy mix. Fluctuations in electricity prices can impact the levelized cost of renewable energy generation, affecting investment decisions and deployment strategies. Research into electricity prices can provide insights into the economic viability of renewable energy projects and inform policies aimed at promoting their adoption.

A "Universal Power Grid" aims to create an interconnected network that seamlessly integrates various sources of renewable energy, traditional power sources, and supporting technologies to ensure a reliable and efficient electricity supply on a global scale. Firstly, in the domain of Reliability and Resiliency, the establishment of a universal power grid necessitates a thorough investigation into its impact on grid reliability and resilience. Integrating diverse sources of energy, including renewables and energy storage systems, introduces complexities in grid operation and management. Research is needed to assess how the universal power grid affects the reliability of electricity supply, particularly in managing uncertainties and faults inherent in renewables, and to develop strategies for enhancing grid resilience in the face of dynamic conditions.

Secondly, supporting technologies such as energy storage systems, synchronous condensers, and communication infrastructure play a vital role in ensuring the efficient operation of a universal power grid. Investigating converter technologies, communicationless control strategies, and optimal grid architecture is essential for seamless integration and operation. Moreover, understanding the impact of the universal power grid on stability, protection, and fault-tolerant control systems is crucial for maintaining grid reliability and resilience, aligning with the research gaps identified in Energy Storage and Real Costs of Renewables.

Furthermore, the concept of a universal power grid correlates with the research gaps in Zero Emissions Energy System Planning and Circular Economy. Developing a comprehensive framework for a universal power grid requires consideration of its environmental impact and sustainability. Research is needed to assess the lifecycle emissions and environmental footprint associated with the deployment and operation of the grid infrastructure, as well as strategies for mitigating these impacts and maximizing resource efficiency, aligning with the objectives of Zero Emissions Energy System Planning and Circular Economy.

The research gap of "Space Energy Transfer" impacts on the broader spectrum of research gaps identified in this book within the energy landscape. Firstly, space-based energy systems offer a promising avenue for addressing the intermittency issues inherent in renewable energy sources like solar and wind power. Therefore, the research gap in space energy transfer intersects with other critical areas such as reliability and resiliency, energy storage, and international grid integration. For instance, leveraging space-based energy systems to enhance grid stability and reliability requires addressing concerns related to electromagnetic interference and the seamless integration of energy transmission into existing infrastructure. Additionally, exploring the potential role of space energy transfer in international grid projects necessitates a comprehensive evaluation framework that considers economic, environmental, and regulatory factors across multiple jurisdictions.

The research gap concerning "Assessing New Projects in Energy Systems" plays a critical role in shaping the trajectory of research efforts across various domains within the energy sector. Firstly, the establishment of a comprehensive assessment framework directly influences the reliability and resiliency of energy systems. By integrating criteria such as technical feasibility, economic viability, and environmental impact into project evaluations, decision-makers can prioritize projects that contribute to enhancing grid reliability and resilience. Moreover, addressing the research gap in assessing new energy projects has significant implications for the development and integration of supporting technologies, such as energy storage systems and communication infrastructure. A robust assessment framework enables stakeholders to accurately evaluate the economic viability and technical feasibility of integrating these technologies into energy systems.

Additionally, by incorporating lifecycle environmental impact assessments into project evaluations, stakeholders can gain insights into the environmental implications of renewable energy projects, facilitating informed decision-making and resource allocation. Furthermore, the development of comprehensive assessment frameworks fosters synergy between different research domains, enabling stakeholders to address multifaceted challenges and optimize the sustainability and efficiency of energy systems.

The research gap regarding the real costs of renewables is foundational and interlinked with various other research gaps identified in the energy sector. Firstly, in the domain of Reliability and Resiliency, understanding the real costs of renewables is crucial for assessing the marginal cost of reliability, particularly in electricity markets where renewables play a significant role. In the context of Energy Storage, understanding the real costs of renewables is essential for effective integration and control of energy storage systems. Research on energy storage strategies must account for the intermittency of renewables and the supporting technologies required to mitigate this variability.

Furthermore, in the realm of Zero Emissions Energy System Planning, assessing the approach proposed in the real costs of renewables is critical for developing sustainable energy systems. Research on the lifecycle of environmental emissions and pollution associated with renewables informs decision-making on achieving zero emissions targets. Understanding the environmental impact of renewable technologies guides efforts to mitigate their footprint and ensure the sustainability of energy transitions.

Moreover, in Electricity Prices and Universal Power Grid research areas, taking the approach of understanding the real costs of renewables is pivotal for designing cost-effective grid solutions. Understanding the cost implications of integrating renewables into electricity markets informs pricing mechanisms and grid redesign strategies. Additionally, in the domain of Universal Power Grid, research on the real costs of renewables contributes to the development of optimal grid architectures and communication-less control strategies, ensuring reliable and efficient operation.

Future Works

Engineering education is at a pivotal juncture to address one of the most pressing global challenges—ensuring universal access to sustainable, affordable, and reliable electricity. By addressing the UN Sustainable Development Goal 7, affordable and clean energy, the world not only shifts from traditional baseload power to a mix of technologies and distribution but also provides an opportunity to give energy to those who have no access today. To do so, the next generation of engineers needs to adapt and transition across various technologies, innovating and engaging in interdisciplinary teams. Furthermore, to drive efficiencies, they will need to learn to use generative artificial intelligence (genAI) responsibly and ethically as a tool. To address this challenge, some industry research questions need answering. (1) Pedagogical Effectiveness: What teaching methods most effectively convey the complex interdependencies of technical, social, and economic factors in sustainable energy solutions? (2) Curriculum Development: How can curricula be structured to balance technical depth with broad interdisciplinary exposure? (3) Technological Integration: How do we effectively integrate GenAI, virtual reality, and simulation tools together with real-world experiences to enhance learning about complex energy systems and their management? (4) Global and Cultural Contexts: How do we ensure educational experiences and outcomes consider the differences across cultural, environmental and economic contexts when teaching about sustainable energy? (5) Assessment: How can learning outcomes in the cognitive, psychomotor, and affective domains be effectively assessed in interdisciplinary energy education? (6) Reflection: What metrics best measure the impact of educational interventions on students' readiness to tackle real-world energy challenges?

Some areas are identified in this book as research gaps, but are not analyzed in detail here. For example, in nanotechnology, there remains a need for further exploration into the development and application of nanomaterials for enhancing the efficiency and performance of energy storage devices and power generation systems. Additionally, leveraging big data analytics presents an opportunity to optimize energy distribution networks, improve demand forecasting, and enhance operational efficiency, yet there is a lack of comprehensive studies addressing its full potential within the electricity sector.

Implementing communication between grid-connected inverters presents several challenges, and addressing these challenges requires collaboration among stakeholders, including manufacturers, utilities, regulators, and technology providers. (1) Interoperability: Inverters from different manufacturers may use proprietary communication protocols or standards, leading to compatibility issues. To address this, standardization efforts are crucial, focusing on the development of unified communication standards tailored for inverters. (2) Scalability: As the UPG expands to accommodate additional renewable energy sources and distributed generation, the communication infrastructure must scale accordingly to support a growing number of inverters. To achieve this, scalable architecture designs and optimized communication protocols for scalability are essential. (3) Reliability: Communication failures or disruptions can adversely affect grid stability and performance. Ensuring reliable communication between inverters, especially in harsh environmental conditions or remote locations, requires robust communication protocols and fault-tolerant systems. (4) Cybersecurity: Inverters are potential targets for cyberattacks, which can disrupt grid operations, compromise data integrity, and pose safety risks. Implementing robust cybersecurity measures, such as encryption, authentication, and intrusion detection systems, is essential to protect communication networks and ensure the integrity of grid control systems. A promising avenue for research involves leveraging wireless communication, edge computing, and artificial intelligence to enhance communication efficiency, reduce latency, and improve scalability in microgrid communication systems.

A key research question with a wide range of rectennas is how to maintain high gain in multiple directions and hence increase the amount of collected energy. A significant number of works are focused on powering and recharging implanted medical devices. This cover frequency ranges from 900 MHz to 1.5 GHz. We envisage that this area will grow further as implantable and other forms of non-invasive surgery become more common place. A recent work [1] has provided a comprehensive survey of RF energy harvesting. The book includes a review of 12 previous surveys and goes on to further 130 contributions in the area of RF energy harvesting and wireless power transfer. The focus is on short-range energy transfer in the order of a few to 10s of meters. There is significant research addressing multidirectional energy harvesting, where the direction of the source is not well identified. Therefore, key future research directions could be in the area of flexible materials and high-density energy storage, high-efficiency transfer, and multi-band and multi-direction energy harvesting.

Furthermore, accurately assessing the real cost of electricity, including both direct and indirect expenses related to production, distribution, and environmental impacts, is essential for informing policy decisions and promoting the adoption of sustainable energy solutions. Addressing these research gaps will be critical in advancing the goals of sustainability, affordability, and reliability in the electricity sector.

Detailed research is also needed to investigate affordable, reliable, and sustainable water systems that provide quality energy sources. This includes exploring energy-efficient technologies for water treatment and distribution, renewable energy solutions to power water infrastructure, and innovative methods for water recycling and conservation. By integrating renewable energy with advanced water management practices, we can ensure a steady supply of clean water while minimizing costs and environmental impact. This research will not only improve water access in underserved communities but also support the broader goal of sustainable development.

Conclusion

In conclusion, this book has underscored the multifaceted nature of achieving sustainable, affordable, and reliable electricity, in order to make electricity available to everyone around the world with the perspectives on generation, delivery, operation, and consumption. The identified 15 topics present critical areas where further investigation and innovation are imperative to advance the global energy transition toward a more sustainable future. From enhancing reliability and resiliency to integrating electric vehicles and energy storage solutions, addressing the complexities of zero-emissions energy system planning, and embracing circular economy principles, each gap represents an opportunity to reshape the energy landscape for the better. Furthermore, considerations such as the role of nuclear power, electricity pricing mechanisms, the development of universal power grids, and the exploration of space energy transfer mechanisms are essential in shaping the future trajectory of energy systems. Moreover, the significance of international collaborations and the need for robust assessment frameworks for new energy projects cannot be overstated. Additionally, understanding the real costs of electricity, harnessing the potential of nanotechnology, optimizing communication and security protocols, and leveraging big data analytics are pivotal in navigating the evolving energy landscape effectively. As we embark on this journey towards Threeable electricity, it is imperative that stakeholders from across academia, industry, government, and civil society collaborate synergistically to address these research gaps and advance innovative solutions that ensure universal access to sustainable, affordable, and reliable electricity for all. Through concerted efforts and collective action, we can usher in a new era of energy resilience, equity, and prosperity for generations to come.

Reference

1. T. E. S. Agency. (2023). *Solaris*. Available: https://www.esa.int/Enabling_Support/Space_Engineering_Technology/SOLARIS.

MIX
Papier aus verantwortungsvollen Quellen
Paper from responsible sources
FSC® C105338

If you have any concerns about our products,
you can contact us on
ProductSafety@springernature.com

In case Publisher is established outside the EU,
the EU authorized representative is:
Springer Nature Customer Service Center GmbH
Europaplatz 3, 69115 Heidelberg, Germany

Printed by Libri Plureos GmbH
in Hamburg, Germany